Advance Praise for *Autocracy 2.0*

"As China experiences its DeepSeek AI moment and leads the world in electric vehicles and other green technologies, Westerners can no longer be smug about their technological primacy. This illuminating book explains how authoritarian China accomplished these miracles despite heavy-handed social control and Western sanctions."
—Susan L. Shirk, author of *Overreach*

"*Autocracy 2.0* reshapes the debate on the threat posed by China. It convincingly demonstrates that China will innovate and compete successfully in the twenty-first century. A sobering must-read for anyone interested in American foreign policy and international relations."
—Dan Reiter, author of *How Wars End*

"A bold and well-argued attack on the thesis that authoritarian regimes are technologically stagnant. Lind marshals considerable evidence to show how China has moderated its authoritarian viciousness in order to become a top innovator in multiple high-tech industries. A must-read for political scientists, growth economists, and China-watchers."
—Mark Zachary Taylor, author of *The Politics of Innovation*

"A risky line of thought suggests China is doomed to slow down, with benign geostrategic implications. Lind argues that resting on democratic laurels is not a safe bet. *Autocracy 2.0* integrates grand strategy with political economy to show how China's 'smart authoritarianism' works."
—Stephan Haggard, University of California San Diego

"One doesn't have to agree with the whole book to accept its one big insight: China is very smart. Americans underestimate it at their peril. This book will open their eyes to the remarkable sophistication of Chinese elites. A must-read."
—Kishore Mahbubani, author of *Has China Won?*

Autocracy 2.0

A VOLUME IN THE SERIES

Cornell Studies in Security Affairs

Edited by Austin Carson, Alexander B. Downes, Elizabeth N. Saunders,
Paul Staniland, and Caitlin Talmadge

Founding Series Editors: Robert J. Art, Robert Jervis, and Stephen M. Walt

Autocracy 2.0

How China's Rise
Reinvented Tyranny

JENNIFER LIND

Cornell University Press

Ithaca and London

For Daryl, with love and appreciation

First published 2025 by Cornell University Press

Librarians: A CIP catalog record for this book is available from the Library of Congress.

ISBN 9781501784149 (hardcover)
ISBN 9781501784156 (pdf)
ISBN 9781501784163 (epub)

GPSR EU contact: Sam Thornton, Mare Nostrum Group B.V., Mauritskade 21D, 1091 GC, Amsterdam, NL, gpsr@mare-nostrum.co.uk.

Contents

Contents

Preface

It was the 1980s and Japan was going to become, in the words of a buzzy book at the time, Number One. Articles described how Japanese workers did calisthenics in the morning before they applied themselves briskly to their jobs of taking over the world economy. Pundits and politicians lamented how Japan Inc. stole American technology and how its mercantilist economy was rigged against the United States.

A trip to Japan with my high-school jazz band gave me the chance to experience this rising power firsthand. I was dazzled. I studied Japanese in college and worked in Japan after I graduated. Eventually Japan's bubble economy burst, but I continued to study Japan and through it found my career in US foreign policy and East Asian international relations.

In the 2000s, I started to hear a familiar refrain. China was experiencing rapid economic growth, and Americans began having one of their recurrent debates about US decline. Bookstore shelves turned scarlet, filled with titles predicting China's imminent collapse or inevitable dominance. Looking at those titles, I couldn't help but think of similar books during Japan's rise: *Trading Places*, *The Coming War with Japan*, and the prescient *The Sun Also Sets*. In 1987 Paul Kennedy published his masterful *Rise and Fall of the Great Powers*; on the cover a drawing of a globe depicts Uncle Sam stepping off the top to follow John Bull downward, while a bespectacled Japanese businessman ascends. And before Japan there was the Soviet Union. "A man is running against us but we are not running against him," noted a 1953 *Foreign Affairs* article about Soviet economic growth. "We are merely out for our usual constitutional." The article warned, "One certain fact emerges: the man is going faster than we are. If the race is an infinite marathon it is thus logically inevitable that we shall lose."

I wondered, Why are we so bad at this? How is it that we still can't figure out what leads a country to successfully rise to great power? This question motivated this book project.

As I dug into the topic, it became clear that for China, the key to its great-power potential was technology. With a massive population, a huge economy, and growing military power, China had many of the ingredients of a great power. But great powers must also excel technologically: Technological leadership propels their economies forward and confers important advantages in warfighting.

As observers debated China's rise, however, many argued that authoritarian China could never innovate. The internet giggled at memes about counterfeit Chinese products ("Sunbucks" coffee, sneakers with brands called "nkie" and "daiads," and so on). This sounded a whole lot like what Americans said about Japan after World War II. After all of the hot takes about copycat Japan—hierarchical Confucian culture! rote memorization in education!—Japan became, and remains, one of the world's most innovative countries. Might China also defy expectations? This book is my answer.

This project was a long journey, enriched by the insights of many students and colleagues. At Dartmouth I live in an intellectual stronghold of bullishness about unipolarity's longevity and skepticism about Chinese power (both of which this book challenges). I am especially grateful to my Government department colleagues Stephen Brooks and William Wohlforth, and former Dickey Center postdoc Michael Beckley, who provided invaluable advice and tremendous intellectual camaraderie. I deeply appreciate the many colleagues who read and provided feedback on the manuscript. In addition to Bill and Steve, I thank Ben Valentino, Mike Mastanduno, Jason Lyall, Yusaku Horiuchi, Nicholas Miller, and other members of the Dartmouth-Dickey extended family, including Dan Reiter and Joshua Shifrinson. Ellen Frost and Kenneth Sharpe offered valuable insights and steady encouragement over the years. I am grateful to Tom Candon, Victoria Holt, and Dan Benjamin for fostering a vibrant intellectual community at the Davidson Institute for Global Studies in the Dickey Center, which tremendously enriched this project. Finally, I very much appreciate the contributions of Dartmouth students who took classes that I developed from this research (the Rise and Fall of Great Powers, the Rise of China, and Dictator 101). Our conversations made this a better book.

Writing this book brought me into numerous subfields. Dartmouth's economics department has the world's nicest economists, and I'm grateful to James Feyrer, Doug Staiger, and Doug Irwin for their comments and encouragement. As the project turned toward innovation, I benefited tremendously from my MIT training and network. MIT's Barry Posen and Harvey Sapolsky introduced me to innovation studies during graduate school, and during this project Taylor Fravel offered terrific comments and an audience at MIT's Security Studies Program. Furthermore, I relied heavily on the work of Zak Taylor, Dan Breznitz, and Doug Fuller—once fellow MIT graduate students, now respected scholars. I'm also grateful to Andrew Kennedy, Adam Segal, and Matthew Brummer for their feedback and for helping me navigate the innovation literature.

I am blessed with an East Asia posse as large, multinational, and vibrant as its region of interest. Participation in regional dialogues—sponsored by the University of California, San Diego and by Japan's Sasakawa Peace Foundation—exposes me to key regional trends and debates: enhancing my research and teaching immensely. I couldn't have written this book without standing on the shoulders of numerous China experts. I owe particular thanks to Susan Shirk both for her scholarship and her generosity in reading and commenting on the manuscript. I also thank Alastair Iain Johnston and David Shambaugh for their scholarship and encouragement, and the China and Taiwan teams at Brookings, especially Ryan Hass, for deepening my understanding of the semiconductor industry. Middlebury colleagues Jessica Teets, Orion Lewis, and Sebnem Gumuscu also gave me terrific feedback.

I presented this research at several institutions, including Stanford's CISAC, UCSD's 21st Century China Center, Georgetown's International Theory and Research Seminar, Princeton's Center for International Security Studies, MIT's Security Studies Program, and Brown's Watson Institute. I'm grateful for those opportunities to share my work and grateful to the many graduate students and faculty there who engaged with my ideas, especially Scott Sagan, Kathryn Stoner, Oriana Skylar Mastro, Tai Ming Cheung, Stephan Haggard, Victor Shih, Margaret Roberts, Aaron Friedberg, John Ikenberry, Elliot Ji, Ken Oye, Tyler Jost, Ed Steinfeld, Caitlin Talmadge, and Rush Doshi. Many thanks to Øystein Tunsjø for bringing me into his Asia Security working group. That group, and my Chatham House activities—for which I am so thankful to Leslie Vinjamuri for her support and intellectual camaraderie—created valuable opportunities for thinking about China outside the US bubble.

This book is much stronger because of the stellar team at Cornell University Press. I thank my editor, Jacqueline Teoh, for her keen eye and her support for the project and for me as a scholar. I am indebted to Alexander Downes and an anonymous reviewer for valuable feedback. I also thank Martyn Beeny and CUP's marketing team for their creativity and hard work on the book's behalf.

Finally, I am deeply thankful for my family's unwavering support. My father, whom I lost during this project, always believed in me. Cathy and Joe Parisie have cheered me on with unfailing encouragement. (They read every word of my first book, including the footnotes, and I bet they will again.) My children Eleanor and Ian encouraged me at every turn with

Gen-Z words I didn't understand but appreciated nonetheless (apparently this book "eats" and I'm a girlboss).

For always being there for me—across endless conversations in which he helped me figure out this book, and as the most wonderful husband, father, partner, and colleague I could have ever imagined—I dedicate this book to my husband, Daryl Press.

Autocracy 2.0

Introduction

> He should pursue moderateness in life, not the extremes ... he should
> seek the company of the notables, but seek popularity with the many.
> As a result of these things, not only will his rule necessarily be nobler
> and more enviable by the fact that he rules over persons who are
> better and have not been humbled and does so without being hated
> and feared, but his rule will also be longer lasting; further, in terms of
> character he will be in a state ... half-decent: not vicious but half-
> vicious.
>
> —Aristotle, *Politics*

It was the drums. We should have known from the graphs: those lines
shooting steeply skyward. We should have known from the spreadsheets,
from the production statistics, from the Louis Vuitton and SUV sales. Or at
least from all those Su-27 fighter jets. But it was the drums that did it. When
Beijing's Bird's Nest stadium went dark that summer night in 2008, and
when two thousand and eight men pounded their bronze *Fou* drums and
shouted Confucian verse in perfect unison, two billion pairs of eyes wid-
ened all over the world, and finally understood. Commentators effused
about the 2008 Olympics as China's "coming out party": as the moment
China rejoined the ranks of the great powers. Scholars smiled indulgently
at media hyperbole—but the media weren't wrong. Around the time that
those 2,008 men struck their illuminated drums, China had become a great
power.

China's rise has profound consequences for the world. Great powers
fight wars against one another. They compete for territory, spheres of influence, allies. They jockey to set rules and standards in international diplomacy, finance, trade, and technology. The number of great powers matters
too; scholars argue that different distributions of power (e.g., multi-, bi-, or
unipolarity) bring different dangers, and that shifts in the distribution of
power (i.e., power transitions) elevate the risk of war.[1] China's ascent to
great power thus has implications for the stability of international politics,

the character of international order,[2] the competition between democracy and authoritarianism,[3] and the foreign policies of the United States and other countries.[4]

Not so fast, many might protest. A number of influential thinkers argue that the world remains unipolar—that China faces a huge gap in its economic, technological, and military power vis-à-vis the United States.[5] Prominent studies conclude—formidable Chinese drummers notwithstanding—that "the leading and rising states diverge technologically to a dramatic degree," and that China "is at a fundamentally different technological level" than the United States.[6] Skeptics further argue that China would be unable to close this gap. As Andrea Gilli and Mauro Gilli observe, "The increase in the complexity of weapon systems has exponentially raised the absorptive capacity requirements to assimilate foreign know-how and experience and to imitate foreign military platforms."[7]

Others point out that trends in China aren't looking promising. They note that Chinese growth is slowing and faces several "headwinds."[8] China experts further warn that increasingly repressive policies under Xi Jinping will harm China's future economic prospects.[9] Economists argue that China has reached the developmental stage at which it must shift from input-led growth (i.e., driven by labor and capital inputs) to innovation-led growth.[10] But many observers, drawing on an extensive literature about institutions in economic development, contend that China's authoritarian ("extractive") institutions inhibit innovation because they stifle information flows, creativity, and market forces.[11] In this view, China will be unable to sustain its economic growth or develop the cutting-edge military technology it needs to compete against the United States. All told, skeptics see a huge gap in China's science and technology (S&T) capabilities, which the country must close in order to catch up with the United States, and they doubt that the country can succeed because of its (increasingly) authoritarian institutions.

As an East Asia expert who got started in her career with a focus on Japan, I found these debates surrounding China's rise—and its apparent limits—all too familiar. After World War II, while debating how to handle Japanese economic reconstruction, many US officials were dismissive of the idea that Japan could ever produce anything that anyone wanted to buy. As Americans had watched the rapid rise of Meiji-era Japan, "Western propaganda," noted George Packard, "had portrayed . . . Japanese culture as full of imported imitations from China and the West."[12] During Japan's postwar economic "miracle," many observers continued to dismiss Japan as a copycat: as a country that was good at kludging together products innovated by creatives in Boston, Oxford, or Zurich. Similarly, when Taiwan experienced economic growth in the 1970s, Americans equated "Made in Taiwan" with cheap consumer products. We even wrote off South Korea as a producer of tiny cheap cars (and then bought them in the millions). Today, Japan, South

Korea, and Taiwan produce some of the world's most sophisticated technology.

The ubiquity of all-too-familiar "Why China Can't Innovate" claims, plus the importance of China's economic rise for world politics and US foreign policy, motivated this book. Once again, a country was generating rapid catch-up growth, and once again, many were dismissing it as a copycat that would be unable to innovate—and thus unable to sustain its economic growth. But this time was different. This country was not a US ally. It was notably unhappy with many aspects of the post–World War II US-led order. And given its massive size, it was a potentially formidable great-power competitor. We couldn't, and can't, afford to get this one wrong.

Smart Authoritarians

This book draws on arguments about economic development, innovation, and authoritarian politics to make two key claims. First, I show that China has already become one of the world's most technologically advanced countries. Its fintech companies have innovated highly efficient digital payment systems, transforming financial convenience and inclusivity. In artificial intelligence (AI), Chinese firms perform at the leading edge in facial recognition, autonomous driving, and natural language processing. In telecommunications China leads the development and implementation of 5G networks, expanding connectivity and enabling the Internet of Things (IoT) revolution. The global leader in green technology, China produces a substantial portion of the world's solar panels, wind turbines, and electric vehicles. China has also established itself as one of the world's most elite space powers, launching a space station, landing a rover on Mars, and retrieving samples from the far side of the moon. This book draws upon several metrics from the innovation literature to show how China has—recently and dramatically—improved its S&T capabilities. China's excellence in emerging technologies suggests that this strength will continue in the nascent "Fourth Industrial Revolution."[13]

To be sure, China's national innovation system has weaknesses to overcome, and the country now faces a less supportive international climate.[14] China is not the world's most innovative country; across several metrics, the United States continues to dominate. But China need not overtake the United States—economically, militarily, or technologically—in order to challenge it in a dangerous security competition. Rather, the question is whether China has joined the United States and other countries at the global technological frontier. Metrics of innovation show that it has.

China's technological excellence raises a puzzle. If the "extractive" institutions that authoritarian leaders rely on to stay in power undermine innovation, what explains China's S&T success? My second claim is that China's government achieved this through what I call "smart authoritarianism." Arguments about a democratic advantage characterize authoritarian regimes as homogenous and static, when in fact they are heterogenous and adaptable.[15] In response to the advent of the information age and other late-twentieth-century trends, many authoritarian leaders recognized a growing tradeoff between the extractive institutions that kept them in power and the conditions necessary to foster growth and innovation. Smart authoritarians adapted by pursuing more inclusive economic policies and governing through low-intensity repression.[16] They became, as Aristotle advised autocrats, "not vicious but half-vicious."

China's future competitiveness depends on its ability to remain at the technological frontier—namely, on the Chinese Communist Party's (CCP's) ability to continue to manage the tensions between extractive policies of authoritarian control, and more inclusive policies required to stimulate innovation. While China's future success is uncertain, this book explains how China—contrary to the pervasive conventional wisdom that authoritarian regimes cannot provide the conditions to foster innovation-based growth—rose to become a superpower and a technological leader. China, as I show, tempered its authoritarian viciousness not only to lift itself out of poverty but also to power its startling technological ascent.

Why This Matters

Grappling with China's rise to become a peer competitor of the United States—and how it did so—is critical for key policy initiatives and multiple scholarly literatures. First, China's rise to great power suggests a return to a hypercompetitive bipolar world.[17] China, now a superpower, will likely engage in common superpower behaviors. It will likely seek regional hegemony: undermining US alliances, interfering in its neighbors' politics to encourage pro-China policies, and seeking to control territories it lost in the nineteenth century.[18] Beijing is already trying to influence norms and standards within the international system in ways that favor its own firms and political system.[19] China will continue to support authoritarian leaders and slow the spread of democracy around the world.[20] Smart authoritarianism—which shows that authoritarian countries can be more adaptable and competent than many observers currently believe—represents a grave threat to global democracy.[21]

Findings from this book also contribute to debates about China's future. Many China-watchers, particularly since the COVID-19 shock, have argued that China's economy will struggle to grow in the future—that China has "peaked."[22] China experts also describe a range of Xi's policies as economically harmful.[23] In this view, Xi is behaving irrationally: At a stage when China must innovate, Xi is shooting himself in the foot by enacting policies that quash innovation. The smart authoritarian view, however, would inter-pret Xi's move toward greater control as a prudent correction to years of more inclusive policies, and the societal forces that those unleashed. According to the smart authoritarian model, the CCP should loosen up when conditions allow it, as it has done in its evolving policies toward Chi-na's technology sector.[24]

The key questions for the future are (1) whether Xi's policies, though more repressive relative to previous eras, will still provide the conditions to support innovation, and (2) how these policies will affect different **types** of innovation.[25] No one (including Xi) knows how much repression a modern, globalized, information-age economy can tolerate; furthermore, no one knows whether innovation can coexist (or which types of innovation can co-exist) with the unprecedented technologies of control on which today's autocrats rely.[26] Indeed, some scholars find that investing in technologies of political control (such as AI) boosts a country's domestic industries and global technological competitiveness.[27] Amid all of this uncertainty, by explaining how China achieved levels of growth and innovation that many had deemed impossible, this book lays the groundwork for thinking through—and preparing for—China's future.

In articulating smart authoritarianism, this book joins ongoing conversa-tions about heterogeneity among authoritarian regimes. In the late twenti-eth century, stunning economic growth in East Asia's "tiger" or "miracle" economies led a large comparative political economy literature to study what scholars called developmental authoritarianism. Scholars showed how illiberal regimes in South Korea and Taiwan engineered rapid eco-nomic growth, in stark contrast to authoritarian dysfunction and poverty in countries like the Philippines and Zaire. This book builds upon and con-tributes to that literature by showing that authoritarian countries can do more than preside over catch-up growth. They can also cultivate cutting-edge innovation, something many observers continue to believe is rare to impossible. In doing so, the book challenges a highly prominent multidisci-plinary literature about the relationship between extractive institutions and innovation. My findings suggest that scholars, pundits, and policymakers should move past thinking about democracies and authoritarian regimes in binary terms to explore the effects of heterogeneity among authoritarian regimes, and should also take into account the diversity of innovation types.

Finally, this book has important findings for arguments about a "demo-cratic difference" in a variety of IR phenomena. Much of the thinking about this so-called difference should be reexamined in consideration of the diversity of authoritarian regimes. Ground-breaking work in security stud-ies has already begun to do this, and others should follow their lead to test arguments about how authoritarian heterogeneity affects diplomacy, inter-national political economy, and other areas.[28] Exploring technology and great-power rise, this book contributes to and hopefully accelerates this trend.

What's Next

The remainder of this book proceeds as follows. Chapter 1 sets the stage for the book's claims. I argue that great powers must not only have among the world's strongest economic and military capabilities; they must also be able to compete at the global technological cutting edge. A country's competitiveness against other great powers depends on its ability to innovate in its commercial economy and to employ new technologies on the battlefield. Describing the stages of economic development through which rising powers ascend, I note that innovation is essential for sustaining economic growth past middle income. Certain conditions—property rights, capital and foreign-direct investment, macroeconomic stability, and high-quality human capital—are important in order to foster innovation. Furthermore, innovation results in "creative destruction" that requires ongoing political-economic reform.[29]

Can China provide these conditions? At stake is its competitiveness as a great power—affecting its ability to advance its territorial claims and challenge international order. I lay out the skeptical view, grounded in the institutions literature, which asserts that although authoritarian countries can preside over "catch-up" growth, they cannot foster innovation-based growth. In this view, the extractive institutions that keep dictators in power clash with the inclusive policies—freedom of information and mobility, openness, and so on—that allow innovation to thrive.[30]

Chapter 2 evaluates Chinese S&T capabilities and concludes that, contrary to the skeptical view described in the previous chapter, China has emerged as one of the world's most innovative countries. I define *innovation* as having multiple levels and types, and I summon metrics to measure it at the national level. Such metrics include both inputs (e.g., tertiary education and R&D spending) and outputs (e.g., high-tech exports, patents, and highly cited scientific publications). These indicators show China competing at the technological cutting edge. China is generating the commercial technology that confers economic advantage, and the military technology required for great powers to compete on the battlefield.

Chapter 3 takes up the puzzle posed by China's impressive S&T success: How can a country governed by extractive institutions create the conditions to foster cutting-edge innovation? I argue that "smart authoritarianism" explains China's success in narrowing the economic and technological gap with the United States. Smart authoritarianism agrees with the institutions school that authoritarian countries face tensions between political control and the inclusive conditions necessary to foster innovation. But institutions theory erred in seeing an irreconcilable tradeoff between political control and innovation, and in viewing authoritarian regimes as homogenous and static. Although some authoritarian regimes remained highly extractive (and poor), smart authoritarians adapted, implementing more inclusive—yet still authoritarian—policies in order to provide the conditions to foster catch-up growth and, later, innovation.[31] Leaders refrained from expropriation and created judiciaries to protect property rights; governments invested in human capital; they permitted a (government-controlled) civil society; as countries advanced through stages of economic development, leaders implemented political-economic reforms. Authoritarian regimes are thus a diverse group—and some of them adapted to the changing global economy in order to foster both economic growth and innovation.

Chapter 4 makes the case for China's smart authoritarianism by tracking the arc of China's rise. I begin with its fall: the failure of China's nineteenth-century Qing leaders to compete technologically, which led to military disasters, to the Qing's toppling, and to decades as a failed state. China then experienced years of poverty under Mao Zedong's brutal authoritarian rule. Notably, however, under Mao, more inclusive policies toward women, health, marriage, and education caused a demographic transition, laying the foundation for China's later economic growth.

China's shift to smart authoritarianism began in 1978, under the leadership of Deng Xiaoping. The CCP reformed the Party and civil service, improved property rights, and created stable economic conditions for investment. It shifted from high-intensity repression to low-intensity repression. It later relied on (and became the world's leader in) technology to control speech and information deemed harmful to the Party. The CCP continued to improve educational access and quality. "Reform and Opening Up" enabled China to access markets, FDI, global knowledge networks, and aid from international institutions. Smart authoritarianism fueled not only China's spectacular catch-up growth but also its recent innovation surge.

Authoritarian adaptation is not a new phenomenon—and it is not confined to China. Chapter 5 steps outside the first few chapters' focus on China to consider authoritarian adaptation elsewhere. It first discusses how authoritarian regimes throughout history faced and adapted to changes in technology and the global economy. In the late twentieth century, a leader who pioneered the smart authoritarian model, and helped disseminate it around the world, was Singapore's Lee Kuan Yew. The chapter shows how Lee's authoritarian government engineered a demographic transition to unleash economic growth, raised the level of human capital, created self-constraining institutions and a professionalized civil service, opened its economy to trade and travel, and permitted a flourishing—albeit government-constrained—civil society. These policies lifted Singapore from perilous poverty into rapid catch-up growth, and later into the high-income category.

Lee did more than make his country prosperous; he traveled the world advising other autocrats about how to follow in his footsteps. Chapter 5 follows Lee's travels: From Kazakhstan to the United Arab Emirates to Ethiopia, authoritarian regimes embarked on reforms in order to foster economic growth and promote innovation. Such cases suggest that the model pursued by the tiny city-state may have more portability than the institutions school expects.

China has emerged as a S&T superpower and peer competitor of the United States, and smart authoritarianism explains how. But what does the future hold? Chapter 6 engages with pessimistic arguments about China's economic future. I argue that China's ability to compete with the United States depends on its continued economic growth (although at much-reduced levels), on its continued mobilization of military power, and on its ability to remain at the global technological frontier. Many observers argue that China's growth is slowing due to several adverse trends.[32] China experts also warn that CCP policies are shifting toward greater personalism, statism, and repression—not so smart anymore?—which scholars argue will impede future growth.[33] This concluding chapter uses the smart authoritarian lens to contribute to debates about China's future. China's ability to remain a superpower and to compete against the United States depends, I suggest, on the CCP's continued management of the tensions between the extractive policies of authoritarian control and the inclusive policies that foster innovation. Authoritarian China's future depends, in short, on its ability to stay smart.

The Stakes and the Debate

In Francis Spufford's novel *Red Plenty*, a fictional Nikita Khrushchev visits the United States and regards its economy with wonder. "They were magnificently good at producing things you wanted—either things you knew you wanted, or things you discovered you wanted the moment you knew that they existed. Somehow their managers and designers thought ahead of people's wants." Spufford's Khrushchev concluded that to compete, "Soviet industries would have to learn to anticipate as cleverly, *more* cleverly, if they were to overtake America. . . . They too would have to become experts in everyday desire."[1] But neither Khrushchev (fictional nor real) succeeded in fostering innovation within the Soviet economy—let alone in overtaking the United States. Observers blame the Soviet Union's failure on its domestic institutions.

Decades later, observing China's rise and debating its future, skeptics expressed similar doubts that the authoritarian country could foster the cutting-edge innovation necessary to sustain its economic growth and to develop military technology competitive with that of the world-class US military. This chapter develops these skeptical arguments—which the rest of this book challenges.

I begin the chapter by discussing the role of technology in great-power competition, explaining that great powers must operate at the global technological frontier if they are to successfully compete economically and militarily. I then describe a rising power's ascent through different developmental stages, in which innovation in science and technology (S&T) becomes essential at the middle-income stage. Drawing on an innovation literature, I argue that innovation requires several supporting conditions: for example, property rights, high-skilled human capital, civil society.

What kinds of societies are able to provide these conditions? A prominent argument holds that domestic institutions are key. The institutions school holds that in the modern era, although autocracies can preside over economic growth for a time, *only* democracies provide the conditions that foster innovation, suggesting a significant democratic advantage not only in economic growth but also in great power politics. This chapter thus lays out what's at stake—China's superpower future—and examines a prominent argument skeptical of China's success.

The Stakes: China's Rise and the Balance of Power

China's rise to great power, after decades of US-led unipolarity, is transforming international politics. Great powers compete to create political and economic orders.[2] Great powers fight wars that slaughter millions of people around the world.[3] Wary great powers create spheres and buffers, they cultivate protégés that they support materially and diplomatically,[4] they set up puppet states, and they overthrow regimes in smaller states that are not sufficiently deferential.[5] Scholars argue that transitions in the balance of power create a particular risk of instability and war.[6] China's rise to great power is thus one of the most consequential issues in international politics today.

Great powers are among the world's most materially capable countries.[7] They are massive in terms of **scale**, with the world's largest economies (which require large populations) and the world's largest militaries.[8] With their wealth great powers mobilize large numbers of military personnel, and build large and diverse military forces, enabling them to "project power beyond their borders to conduct offensive as well as defensive military operations."[9] Wealth also enables great powers to fund other instruments of statecraft for great-power competition: a large diplomatic corps, foreign aid, military assistance programs to allies, and so on. A great power's wealth also makes it a model that developing countries want to emulate.[10]

In addition to being massive in scale, great powers must have technological **sophistication**, operating at the global technological cutting edge.[11] As Stephen Brooks and William Wohlforth argue, "Economic capacity is a necessary condition of military power, but it is insufficient; technological prowess is also vital, especially given the nature of modern weaponry. Technological capacity also magnifies economic capability, and military capability also can have spinoffs in both the economic and technology arenas."[12] S&T innovations "create new commercial and industrial sectors (leading sectors) and ways of doing things that are highly significant in propelling industrial and economic growth."[13] Great powers also integrate within their militaries cutting-edge technologies around which they devise innovative tactics.

TECHNOLOGICAL COMPETITIVENESS AND GREAT-POWER COMPETITION

Great-power competition across three (now four) Industrial Revolutions shows that a great power's success depends on its ability to innovate, absorb, and diffuse rapidly changing technologies. The eighteenth century sparked the first Industrial Revolution: "the evolution of the steam engine, the development of iron metallurgy, the shift in fuel from wood to coal, the rise of industrial chemistry, the establishment of a machine industry."[14] These new technologies dramatically raised productivity in the agricultural sector, allowing more food production with less labor. Agriculture shrank as a share of the national economy while the more productive industrial sector grew. Industry adopted the "factory model of production" made possible by the invention of the steam engine and improvements in iron and steel metallurgy. These developments also led to the spread of the railroad, which improved mobility and connected people and markets. Countries that were on the technological cutting edge at that time, notably Britain, experienced a staggering leap in productivity and output, whereas countries that failed to embrace these new technologies—remaining agrarian with small handicraft economies—fell considerably behind in terms of gross domestic product (GDP) and wealth.[15]

Countries that were slow to respond to the Industrial Revolution found themselves disadvantaged militarily. Williamson Murray writes that the first Industrial Revolution made technology and science "crucial to battlefield success," giving a technologically advanced country "an important advantage over its opponents."[16] Navies that sailed iron ships powered by steam demolished wooden ships powered by sail.[17] Countries that laid railroad track and employed steam locomotives could more quickly mobilize and dispatch a larger number of military forces, who arrived at the front healthier and better rested relative to armies on foot.[18] A dramatic evolution in artillery, ordnance, and firearms also made some nineteenth-century armies far more lethal. Writes William Perry, "Soldiers who had these new weapons had an order of magnitude more firepower than those who did not."[19]

Technological backwardness and the resulting military disasters toppled once mighty countries from the great-power ranks. Austria-Hungary's military-technological inferiority led to its disastrous performance in World War I, and to its subsequent collapse.[20] China's failure to adapt to the Industrial Revolution led its economy to be overtaken by the West, and led the country to experience catastrophic military defeats.[21] In the Opium Wars, British troops fielded new flintlock and percussion cap muskets, whereas "the majority of Qing troops relied on traditional weapons such as the bow and arrow, spear, sword, and halberd. Those Qing soldiers who used firearms carried outdated matchlocks, unreliable and sometimes deadly to the user."[22] China's nineteenth-century defeats led to territorial losses, diminished sovereignty, and punishing war indemnities – followed by revolution and decades of chaos as a failed state.

By contrast, technological superiority led to consequential military victories. In the Austro-Prussian War of 1866, Prussians armed with the breech-loading Dreyse needle gun savaged Austrians firing muzzle-loading rifles.

The war delivered impressive gains: "In seven weeks Prussia gained five million inhabitants and 25,000 square miles in Germany."[23] Across the world, Japan's superior naval technology enabled it to defeat Russia in the 1905 Russo-Japanese War, elevating Japan to great-power status and giving it control over the Liaotung Peninsula, southern Sakhalin Island, and Korea. "Quite a colonial haul," observes Max Boot.[24]

The geopolitical importance of technological advantage continued into the twentieth century. The second Industrial Revolution included advances in "electrical generators, internal combustion engines, motor vehicles, airplanes, radios, telephones, [and] radar"; notes Boot. "All of these technologies that contributed to the growth of the world economy between 1919 and 1939 . . . would help to devastate a large portion of the world in the six years that followed."[25] Later in the century, the shift into an information age (the third Industrial Revolution) once again transformed economies, militaries, and the balance of power. The United States and its liberal allies combined advances in electronics and computerization with a globalized production structure to generate dramatic increases in productivity.[26] Key technologies included computers, semiconductors, and later the internet. Although the Soviet Union had been a technological superpower in the second Industrial Revolution, it was unable to keep pace in the Third; the Soviet economy slowed and began to decline in the 1970s.

The United States harnessed its leadership in information technologies to create a revolution in military affairs (RMA) that translated to US dominance in the Cold War and beyond. By integrating computers into weapons and war, the United States transformed conventional warfare: through improved battlefield sensors, greater coordination of operations across the theater, and long-range precision strike. Soviet leaders understood that the information revolution was shifting the military balance of power against them. They realized, wrote Eliot Cohen, "that their country, incapable of manufacturing a satisfactory personal computer, could not possibly keep up in an arms race driven by the information technologies."[27] China, the collapsing USSR, and other countries that had been unable to capitalize on the information revolution watched the United States' RMA—the effects of the information revolution—on full display in the Gulf War of 1990. The USSR's inability to compete contributed to the dissolution of the Soviet Union, and the end of a bipolar international system. At the turn of the century the United States enjoyed a position of geopolitical dominance. Since the first Industrial Revolution, then, a great power's performance at the technological cutting edge has been essential for its geopolitical competitiveness.

STAGES OF GREAT-POWER RISE

Rising powers develop economic might by moving through different developmental stages, in which the role of technological innovation becomes increasingly vital. A country's rise from low income to middle income is described as "catch-up" growth.[28] Catch-up growth first requires a demographic transition (a decrease in mortality and fertility).[29] During this early stage of growth, countries don't produce much indigenous innovation; they experience huge increases in output from mobilizing inputs of labor and capital, and they hoover up technology from countries at the global technological frontier (via imports, theft, and foreign direct investment, or FDI).[30] For this reason, observers often dismiss rising economies as perennial copycats, as people did with Japan in the twentieth century (and as many observers continue to do with China).[31]

Catch-up growth inevitably slows.[32] The supply of rural workers streaming to factories dries up and wages rise, and previously advantageous demographics from a "demographic transition" become less favorable as the workforce ages.[33] Growth also slows as the economy experiences diminishing returns to capital.[34] At the middle-income stage, "Mere increases in inputs, without an increase in the efficiency with which those inputs are used," Paul Krugman explains, "must run into diminishing returns; input-driven growth is inevitably limited."[35] Economists warn of a "middle income trap" in which most economies stall, failing to advance to the high-income category.[36]

Growth past the middle-income level results from productivity gains: from innovation. Following Mark Zachary Taylor, I define *technology* as "a physical product, or a process for physically altering materials, that is used as an aid in problem solving."[37] Technology, Taylor writes, enables people "to perform entirely new activities, or to perform established activities with greater efficiency." Innovation is "the discovery, introduction, and/or development of new technology, or the adaptation of established technology to a new use or to a new environment."[38] After countries reach the middle-income level, as the Organization for Economic Co-operation and Development (OECD) explains, they "produce more output by better combining inputs, owing to new ideas, technological innovations and business models."[39] Firms must create new processes and find new markets, which requires them to understand "the quality, price, and consumer preference points of the global economy, which is a demanding task."[40] As Spufford's Khrushchev noted while admiring the United States, firms must learn to anticipate demand. Homi Kharas and Harinder Kohli note that "marketing, branding, and new product development . . . are the ingredients of innovation and further growth."[41] In sum, in order to continue to rise economically, countries must innovate.

To be clear, the concept of national income and middle-income traps relates to GDP per capita: a concept from development economics that grafts poorly onto discussions of great-power competition. Discussions of China's rise sometimes involve speculation about whether China might fall into the middle-income trap—meaning that it cannot foster the innovation that propels it to the high-income category, so it will see its economic growth stall. Although great powers do indeed need to be among the world's largest and most technologically advanced countries, they don't necessarily need a high per capita GDP.[42] Some great powers (think Imperial Japan of the late 1800s through World War II and the Soviet Union in the 1950s and 1960s) produced high levels of technology and competed hard in great-power politics despite having lower GDPs per capita. The question is not whether China can continue growing into the high-income category such that it will rival the United States. China doesn't need to be a high-income country in order to engage the United States in a dangerous security competition. Rather, the key question is whether China can operate at the technological cutting edge.

What Drives Innovation?

To compete economically and militarily with other powerful countries, great powers face an imperative to operate at the global technological frontier. But what causes a society to be more innovative or less innovative? Although the sources of innovation continue to be actively debated, a large multidisciplinary literature has identified several key drivers of innovation.[43]

First, scholars agree that innovation requires governments to **rectify market failures**. Kenneth Arrow famously argued that technology is "non-rival" and "non-excludable": that new technology spreads easily and has a spillover effect (in other words, a positive externality) for those who did not bear costs and risks in creating it.[44] Inventors and entrepreneurs must be confident that they will be able to recoup the costs of their labor, investment, and risk. Often this is done through intellectual property rights (IPR), often through patents.[45] *Property rights* can be "broadly defined as the rights of a firm or individual to assets, to the revenue streams generated by assets, and to any other contractual obligations due the firm or individual."[46] "Without property rights," argue Daron Acemoglu, Simon Johnson, and James Robinson, "individuals will not have the incentive to invest in physical or human capital or adopt more efficient technologies."[47] Innovation requires not only IPR but also a fair, efficient, and accessible judicial system in which individuals and firms can adjudicate property rights.

Second, innovation requires **capital**: Research must be funded, and firms require investment. Capital should flow to innovative firms (and be withheld from non-innovative firms). Joseph Schumpeter argued that well-developed financial markets allocate capital for its most productive use.[48] Financial markets with diverse actors (foreign and domestic) and diverse investment vehicles—equity, bonds, bank loans, venture capital—are associated with higher productivity and higher per capita income.[49] Governments can also provide capital; basic science is expensive (with high costs in salaries, equipment, and laboratory space). This creates a market failure that governments should address through "public investment in basic research to support the continued emergence of breakthrough innovations."[50] Government investment in infrastructure, including digital infrastructure, encourages innovation "both by facilitating the cheap circulation of disembodied knowledge flows across and within national boundaries, as well as by reducing the transactions costs of international trade and foreign investment flows."[51]

Innovation also requires a **high-skilled workforce**. At the catch-up stage, governments improve productivity by universalizing education at the primary and secondary levels. Innovation-based growth requires further improvements in education quality and access at the tertiary stage: Education, write Kharas and Kohli, "must be re-tuned for a knowledge and innovation economy."[52] High-skilled workers not only invent and innovate, they diffuse innovations throughout different sectors of the economy.[53]

Scholars argue that **civil society** encourages innovation. Civil society encompasses "an intermediate sphere of social organization or association between the basic units of society—families and firms—and the state."[54] It includes, for example, the private business sector, religious organizations, universities, the media, culture and the arts, and organized labor. It serves as a locus of expertise and ideas; as "an important source of information for both citizens and government."[55] Universities are vital nodes; their researchers produce innovation and train the next generation of S&T researchers; they interact with government and industry to encourage the churn of ideas.[56]

Relatedly, **networks, connectivity, and openness** also drive innovation.[57] The "Triple Helix" model emphasizes linkages between universities, the government, and the private sector.[58] A literature on "national innovations systems" highlights the importance of policy, academic, and industry connections, norms, and practices.[59] Argues Taylor, "Clusters are a particular form of network that reduces costs, creates complementarities, and increases spillovers within a concentrated geographic area."[60] Networks also connect via information technologies (e.g., the internet, social media, and collaborative software).

Openness—access to global networks—encourages innovation. Global trade, financial, and research networks diffuse new technologies and ideas; they connect "entrepreneurs and investors with foreign S&T business opportunities."[61] Scholars note that research in the most innovative countries "has tended toward greater cross-border collaboration": that "open economies product the most high-impact research."[62] Scientists and researchers benefit from connecting with people around the world who are working at the technological frontier.[63]

15

The importance of connectivity and openness is suggested by a prominent case of their absence: the Soviet Union. Despite cultivating very high levels of human capital, poor internal and external communications impeded Soviet innovation. "Soviet scientists are often isolated from Western scientific developments. Russian journals are slow to pick up Western discoveries. Some reports are censored," a Western journalist observed in 1984. "Communication problems extend to within the country, too: A Soviet scientist toiling in one area may not be aware of a countryman doing similar work elsewhere."[64]

Political scientists argue that shifting to innovation-based growth often requires extensive **reforms**, in, for example, trade, education, corporate governance, labor regulations, financial markets, and in the country's overall growth model. Because investment inputs will confront diminishing returns to capital, governments reliant on an investment-led growth model must reform. Capital should flow to the most innovative firms (and should be denied to less innovative firms), which suggests the market—not the government—should allocate capital.[65] This may be a significant departure from the model pursued during catch-up growth, in which governments often direct capital to specific sectors or even specific firms. More broadly, in Joseph Schumpeter's logic of "creative destruction," innovation extinguishes old ways and creates new ones: new products, jobs, fields, sectors, and new demands for government regulation and standards. Scholars warn that, as such, innovation will provoke resistance from stakeholders with a concentrated interest in the status quo. Mancur Olson argues that "distributional coalitions" form to defend access to rents (e.g., subsidies, licenses, access to loans, barriers to entry, and trade protections).[66] The failure of the vast majority of countries to successfully shift to innovation-based growth suggests how powerful such stakeholders are.[67]

Table 1.1 summarizes conditions that encourage innovation.

Table 1.1. Conditions that support innovation

Requirement	Mechanism for promoting innovation
Rectification of market failure	Property rights and a well-functioning judiciary assure inventors and entrepreneurs that they will be able to profit
Financial capital	Funds individual entrepreneurs and whole firms
High-quality human capital	Higher education, high-quality health care, and improved infrastructure
Civil society	Encourages and disseminates information and ideas; universities train human capital
Networks/ mobility/ openness	Connect researchers, investors, firms, and customers; disseminate information and ideas; diffuse innovation across firms and sectors; enable access to foreign markets and foreign direct investment (FDI)
Reform	Supports new firms, products, and sectors while eliminating less innovative ones

Innovation and Domestic Institutions

A prominent school of thought holds that democracies are better able to create the conditions described above (e.g., intellectual property rights, ample capital, a vibrant civil society). Such arguments are nested within a massive literature on domestic institutions and economic development, which dates back to Enlightenment thinking about the virtues of liberal republics.[68]

What are "institutions"? Scholars define *institutions* as "a set of humanly devised behavioral rules that govern and shape the interactions of human beings, in part by helping them to form expectations of what other people will do."[69] Douglass North characterizes institutions as "the rules of the game" in a society, which "structure incentives in human exchange, whether political, social, or economic."[70] Institutions can be formal (e.g., regulations, laws, treaties) or informal (norms, traditions, rituals). Political institutions establish norms, rules, and rights for political participation and access to public office. Such institutions govern how a people choose, monitor, and evaluate their government—whether the people can hold it accountable and "throw the bastards out." Economic institutions govern commerce: As Dani Rodrik writes, "every successful market economy is overseen by a panoply of regulatory institutions, regulating conduct in goods, services, labor, asset, and financial markets."[71]

Scholars situate institutions on a continuum; at one end are "open-access orders" or **inclusive** institutions.[72] These provide people with access and opportunity, connecting them to the government, economy, and society. Inclusive political institutions enable people to stand for office, choose leaders through competitive elections, and monitor government performance. Inclusive economic institutions connect the diffuse populace to the economy, by giving them choice and tools so they can participate effectively.[73]

Toward the other end of the spectrum are "limited-access orders" or **extractive** institutions. These confer private access and goods on a minority of people, benefiting the few at the expense of the many.[74] Extractive political institutions inhibit evaluation of the government and confer the ability to choose (and access) leaders on a select few. Extractive economic institutions deny people choice and access to the economy.

A DEMOCRATIC ADVANTAGE IN ECONOMIC GROWTH?

A prominent, multidisciplinary institutions literature argues that liberal societies enjoy an advantage in economic growth. Scholars observe a correlation between GDP per capita and democracy, pointing out that almost all of the world's high-income countries are democratic.[75]

Scholars identify different mechanisms through which democracies better cultivate growth.[76] Democracies are said to better provide and (through

17

independent judiciaries) adjudicate property rights, whereas autocrats resist creating rival branches of government that reduce their power.[77] In contrast to autocratic leaders (who provide "private goods" to a narrow "selectorate"), democratic leaders are said to be incentivized to provide public goods such as education and health care.[78] Democratic governments are said to better moderate consumption, pay off debt, and pursue other policies that encourage economic stability, thus encouraging investment.[79] By contrast, dictators spend lavishly on patronage, default on debt, interfere with monetary policy, and create multiple exchange rates as rents for their supporters.[80] All of these practices are economically devastating: distorting trade, deterring investors, and depriving the economy of vital capital and technology transfer.[81] Scholars have also argued that democracies better implement the reforms that economic development requires. More responsive to the public interest, democracies are expected to be "more willing to implement reforms, which destroy monopolies in favor of the general interests."[82] Olson argues that free societies are more likely to avoid an economy's tendency to shift from "productive" to "redistributive" activities.[83] Acemoglu et al. agree that democracies are "more likely to enact economic reforms that would otherwise be resisted by politically powerful actors."[84]

But arguments about a democratic advantage in economic growth suffer from numerous problems. Critics argue that democracy doesn't correlate with growth,[85] and other scholars don't find evidence for the various theorized mechanisms through which democracies are said to better stimulate economic development. That is, scholars find no difference between democracies and autocracies in terms of public goods provision,[86] trade openness,[87] and ability to attract capital.[88] Political scientists find that autocracies are as likely as democracies to implement political-economic reforms.[89] In the late twentieth century, a large literature on "developmental authoritarianism" studied countries such as Singapore, South Korea, and Taiwan—all led by authoritarian regimes—that logged sky-high growth rates and raised their GDPs per capita from among the world's poorest to the world's richest.[90] They invested in human capital, avoided expropriation, avoided overconsumption and trade-distorting policies, created meritocratic bureaucracies,[91] and engaged in other progrowth policies.

A DEMOCRATIC ADVANTAGE IN INNOVATION?

Institutions theorists counter that although autocracies can sometimes succeed at the catch-up stage, they will be unable to cultivate the innovation necessary for sustained growth. After all, most modern authoritarian catch-up growth either stalled (e.g., Brazil and Chile) or led to political liberalization (e.g., South Korea, Taiwan). The most prominent case of a wealthy, innovative autocracy is Singapore, a tiny city-state that seems an unlikely model.

THE WILD GARDEN VERSUS THE DESPOTIC LEVIATHAN

Scholars do not expect autocracies to be able to provide the conditions (described earlier) that nurture innovation. To be clear, many such arguments are refuted by the experience of the developmental authoritarians, which provided public goods, maintained stable macroeconomic conditions, and so forth. However, scholars arguing for a democratic advantage highlight other drivers of innovation that they argue authoritarian regimes cannot provide.

According to this view, only democracies provide the freedoms and vibrant civil society that encourage innovation. In democracies, people are free and empowered to pursue their chosen profession in which they have the most to contribute. Information flows freely, inside and outside the country.[92] Karl Popper famously wrote about the virtues of an "open society"—with individual liberties and free institutions creating a marketplace of ideas—that encourages scientific progress and "sets free the critical powers of man."[93] Vannevar Bush, the director of the US Office of Scientific Research and Development during World War II, emphasized the need for free inquiry in science. "Scientific progress on a broad front," he argued, "results from the free play of free intellects, working on subjects of their own choice, in the manner dictated by their curiosity for exploration of the unknown."[94] L. Rafael Reif, former president of the Massachusetts Institute of Technology, referenced Bush's vision of innovation "as a kind of wild garden: individuals seeding ideas based on their intellectual interests with no overall design," a vision that seems incompatible with authoritarian diktat and centralized institutions.[95] Furthermore, a free civil society—for example, with independent universities, nongovernmental institutions, an evaluative free press—creates a churn of ideas and facilitates reform.[96]

Many scholars argue that in autocracies, repression of freedoms and an absence of civil society inhibit innovation. Dictators repress freedoms to block antiregime speech and assembly.[97] They create vertical information flows (and impede horizontal channels of communication);[98] they prohibit or tightly control civil society to inhibit evaluation and the emergence of potential rivals. Authoritarian controls of tertiary education interfere with this vital sector. Regime-controlled curricula and propaganda stifle creative thinking and create "no-go" areas for study and research.[99] Such policies not only dampen creativity and initiative among local researchers; they also repel the world's top scholars and scientists, who will choose to work in liberal countries. All of this undermines innovation: "Knowledge societies cannot function effectively without highly educated publics that have become increasingly accustomed to thinking for themselves."[100] Innovation is "much harder to usher in under the stern gaze of the Despotic Leviathan," write Acemoglu and Robinson:

Innovation needs creativity and creativity needs liberty—individuals need to act fearlessly, experiment, and chart their own paths with their own ideas, even if this is not what others would like to see. This is hard to sustain under despotism. Opportunities aren't open to everybody when one group dominates the rest of society, nor is there much tolerance for different paths and experiments in a society without liberty.[101]

Scholars also argue that democracies foster innovation better because they provide connectivity, mobility, and openness. Democracies allow free movement inside and outside their borders, which facilitates the formation of regional clusters and connectivity with global business and knowledge networks.[102] People in free societies shift into and out of jobs that suit their talents; firms move in and out of sectors (and create new ones). By contrast, authoritarian regimes often control mobility in order to subdue the population.[103] Writes Ian Bremmer, "As the citizens of closed states learn more about life beyond their borders and discover that they don't have to live as they do, tyrants must expend more and more effort to isolate their societies."[104] So autocrats control internal mobility and limit the ability of businesspeople, students, and scientists to study and work overseas. Without access to cutting-edge scientific and education networks around the world, isolated autocracies are expected to face a significant disadvantage.

INNOVATION SINCE THE INFORMATION AGE

Arguments about innovation and democracy have a temporal dimension. Before the late twentieth century, authoritarian countries cultivated cutting-edge innovation. Of course, until recently, *all* innovation occurred in authoritarian societies. Nineteenth-century Germany became an S&T powerhouse, building industries around key technologies such as chemicals, optics, electricity, and the internal combustion engine.[105] In World War II, Imperial Japan, Nazi Germany, and the Soviet Union wielded world-class military technology.[106] After World War II, Soviet science was the envy of the world—for example, in mathematics, physics, astronomy, and chemistry.[107] In the 1950s, Soviet scientists and scholars won numerous global awards, such as the Fields Medal and the Nobel Prize. The USSR detonated an atomic bomb and then a hydrogen bomb, and in 1957 successfully tested the world's first intercontinental ballistic missile and first artificial satellite (Sputnik 1). In 1961, Yuri Gagarin became the first person to orbit the earth. "For some time the United States lagged behind us," bragged Soviet leader Nikita Khrushchev (the real one). "We were exploring space with our Sputniks. People all over the world recognised our success. Most admired us; the Americans were jealous."[108] But in the latter part of the twentieth century, the USSR and its authoritarian partners experienced slowing growth, and slid off the global technology frontier. What happened?

Observers argue that several late-twentieth-century trends transformed both the nature of innovation and the structure of the global economy in ways that disadvantage autocracies. First, starting in the 1960s, the information technologies of the third Industrial Revolution advantaged societies that allowed the unfettered flow of information.[109] Second, the rising significance of FDI advantaged liberal and open societies: those whose institutions reassured investors of property rights, and those that permitted the movement of their people around the world for education, business, and tourism.[110]

Third, another key trend was the rise of the liberal development regime (e.g., the Bretton Woods institutions and regional development banks created after World War II). Private investors and liberal gatekeepers at multilateral development banks condemned authoritarian repression, and imposed aid conditionality related to governance. Corruption, coups, and other domestic failings risked isolating a country from the emergent global development regime.[111] Liberal nongovernmental organizations (NGOs) monitored countries' quality of governance and published global rankings that influenced the disbursement of aid and US political and economic support.[112]

Relatedly, the spread of human rights norms raised the economic costs of using force to stay in power.[113] Human rights abuses—such as the violent suppression of protests or the murder of opposition leaders and journalists—were increasingly visible owing to advances in telecommunications, and these abuses repulsed liberal observers around the world. Steven Levitsky and Lucan Way conclude that "for most governments in most poorer and middle-income countries, the benefits of adopting formal democratic institutions—and the costs of maintaining overtly authoritarian ones—rose considerably in the 1990s."[114]

Another watershed event, which undermined the economic health of many authoritarian countries, was the collapse of the Soviet Union and the end of the Cold War. Authoritarian regimes that had previously enjoyed US or Soviet aid found themselves cut off by their former patrons and searching for new sources of support.[115]

For all these reasons, scholars argue for a fundamental tension between the extractive institutions that maintain authoritarian control, and the freedoms, information, and openness that innovation requires. Samuel Huntington warned of a "king's dilemma" in which monarchs would face pressures to modernize; doing so, however, would undermine the king's hold on power.[116] Regime-based arguments about innovation echo this view, arguing that dictators have to choose between staying in power or the reforms needed to transition to innovation-based growth. As Acemoglu and Robinson argue, "Even though extractive institutions can generate some growth, they will usually not generate sustained growth, and certainly not the type of growth that is accompanied by creative destruction."[117] Way notes, "Economic development frequently threatens dictators by fostering the rise of independent sources of commercial, social, and political power that make it harder for leaders to monopolize control."[118]

Such views have led many observers to argue that despite China's astounding catch-up growth, the country would be unable to successfully shift to innovation-based growth. In China, "Conformist mediocrity is rewarded above unsettling brilliance," argued Odd Arne Westad. "The party and its views hover over everything, and hiring and promotion are decided by patronage rather than by talent."[119] Kerry Brown argued that China is ruled by a "risk averse, highly prescriptive knowledge system" unlikely to shift to one that generates "diverse and challenging new ideas about how to do things differently, design things in a new way, or live in a different fashion—the sort of things innovation involves."[120]

Are skeptics correct to doubt China's ability to shift to innovation-based growth? Must dictators choose between staying in power and providing the conditions that foster innovation? Scholars of innovation have their doubts. Taylor writes that "democracy is neither necessary nor sufficient to explain why some countries are better at S&T than others."[121] He argues, "Institutions determine nothing; they are tools, not causal forces."[122] Matthew Brummer finds that democracy, decentralization, and other institutions-based arguments are all statistically insignificant in explaining national innovation rates over time. "After decades of research," he argues, "these institutions-based explanations have yet to identify the 'right policy mix' because, perhaps, it does not exist, due to the fact that an array of different institutions and policies can solve similar challenges to innovation."[123]

To contribute further to this debate, Chapter 2 examines the case of China, and finds that authoritarian China has become a global technological leader. The remainder of the book explains this puzzle by creating and exploring the model of smart authoritarianism.

The Puzzle of Chinese Innovation

China's technological contributions to the world have been profound. At the 2008 Olympics opening ceremony noted in this book's introduction, dramatic special effects, lavish costumes, and thousands of performers showcased China's "Four Great Inventions": the compass, gunpowder, papermaking, and movable-type printing. As impressive a display as it was, many watching the show may have thought that China hadn't invented anything since the thirteenth century. After all, failing to adapt to the Industrial Revolution, China fell behind the West technologically and economically, leading to military disasters, decades of turmoil, and poverty. Many observers thus held a dim view of modern China's economic and technological potential. "Reverend dullness! hoary ideot!" scoffed Ralph Waldo Emerson: "All she can say to the convocation of nations must be – 'I made the tea.'"[1]

Even after China's rapid economic rise in the late twentieth century, many observers continued to doubt its science and technology (S&T) potential. Prominent international relations (IR) studies argued that China lagged the United States technologically by a wide margin.[2] During catch-up growth, China engaged in rampant intellectual property (IP) theft, through cyberattacks and other forms of industrial espionage. US politicians described this as "the greatest intellectual property theft in human history," said to cost the US economy an estimated $300 billion–$600 billion per year.[3] Chinese internet companies copied US e-commerce and social media companies; China's "Fanfou" "resembled Twitter right down to the baby-blue color branding."[4] Chinese firms brazenly imitated many other products—for example, the Landwind SUV (a near-duplicate of the Range Rover), and Chery's QQ, identical to the Chevy Matiz to the point that the doors of the two cars were interchangeable. China's "Uncle Martian" brand used Under Armour's logo and designs. Sporting your Uncle Martian hoodie, you could sip an espresso in a "Sunbucks" Café while scrolling Fanfou on your "HiPhone."[5] The extent of China's IP theft led many observers to dismiss it as a "copycat nation" incapable of indigenous innovation.[6]

Many skeptics drew on arguments, described in Chapter 1, related to domestic institutions and economic growth, asserting that authoritarian countries cannot innovate because of political repression, censorship, and a lack of key freedoms.[7]

This pessimism about Chinese S&T capabilities contrasts starkly with more recent discussions. A 2024 *Economist* cover proclaimed China an innovation superpower.[8] Headlines in *Harvard Business Review* shifted from "Why China Can't Innovate" (2014) to "China's Innovation Advantage" (2021). Influential tech executives and military officials warned that the United States "could lose the tech contest with China" so needs to "wake up."[9] Many observers worried that China is outpacing the United States in key emerging technologies (e.g., quantum science, biotechnology, green energy, artificial intelligence (AI)). "In some races, it has already become No. 1. In others, on current trajectories, it will overtake the U.S. within the next decade."[10] The Pentagon's 2021 report on Chinese military power pointed to "recent accomplishments in space exploration and other fields," and concluded that "China stands at, or near, the frontier of numerous advanced technologies."[11] US military planners described China as a technological peer competitor.[12]

In debates about China, then, scholars have argued that China lags the United States by decades and (hobbled by authoritarian institutions) would be unable to close this gap; alternatively, increasingly one hears that China has become a technological peer of the United States and is rapidly leaving it behind. Which of these views is closer to reality?

The answer to this question has profound implications for international politics. As Chapter 1 described, across three (now four) industrial revolutions, success in great-power competition has required countries to operate at the technological cutting edge. The Soviet Union's failure to compete in the information age led to its economic decline and collapse.[13] Today, China's ability to challenge the United States in a great-power competition requires China to foster innovation in order to sustain economic growth and to develop and wield cutting-age military weaponry. Chinese inadequacies in S&T would thus reduce its geopolitical competitiveness and help sustain US-led unipolarity.[14]

Introducing metrics from the innovation literature, this chapter measures different dimensions of Chinese innovation to find that China has become a global innovation leader. Metrics show the United States as performing at or near the top of the list of the world's most innovative countries; however, China has caught up to (or overtaken) many countries on this list, such as France, Japan, and South Korea. China has gone from being a copycat—an economy based on assembling stuff that was invented elsewhere—to being an innovative country whose firms are starting to dominate key sectors of the world economy. Chinese excellence in emerging technologies—for example, AI, quantum science, biotechnology—suggests that its technological strength will continue into the "Fourth Industrial Revolution."

Why have so many US and other Western observers underestimated Chinese innovation? People in the United States tend to see innovation as a laissez-faire phenomenon—one in which the state should take a back seat—so are skeptical of the Chinese Communist Party (CCP's) heavy involvement in Chinese S&T. Ironically the US experience actually shows that governments can play a powerful role in fostering innovation. Furthermore, skeptical views of Chinese innovation tend to emphasize one type of innovation (i.e., invention) while neglecting the many other types of innovation at which a country might excel. This chapter reminds us that innovation is a broader phenomenon than many observers recognize. Innovation occurs in many places: in so-called ivory towers, in which labs crowded with costly equipment perform basic science; in firms seeking to shave off a few seconds or a few dollars in their production or distribution process; at the grassroots level, where firms and entrepreneurs create products and apps or introduce them to new functions or markets. China has excelled in what is called process innovation:[15] activities neglected by many observers whose image of innovation is dominated by science-based "invention" activities in which China has been weaker.

What Is Innovation?

Chapter 1 noted Mark Zachary Taylor's definitions of *technology* and *innovation*. *Technology* is "a physical product, or a process for physically altering materials, that is used as an aid in problem solving." Technology enables people, Taylor writes, "to perform entirely new activities, or to perform established activities with greater efficiency." *Innovation* is "the discovery, introduction, and/or development of new technology, or the adaptation of established technology to a new use or to a new environment.[16]

People tend to think of innovation as a new-to-the-world idea—a thought that is reflected in many discussions of China's innovation potential. However, "a new idea for a product or process" is what scholars call invention.[17] By contrast, innovation, as Joseph Schumpeter argued, encompasses a wider range of activities: new methods of production, new sources of supply, the exploitation of new markets, and new methods of organization.[18] Stephen Kline and Nathan Rosenberg describe the profound economic effects of ongoing incremental innovation in the petroleum, automobile, aircraft, and computer industries.[19] Dan Breznitz and Michael Murphree agree: "The acts of innovation that translate into economic growth are to be found at least as much in diffusion, application, incremental, organizational, and process innovation as in the development of completely new technologies and products." For example, they describe how Japan came to dominate the automobile industry "not by novel-product innovation but by developing a superior system of production."[20]

The steel industry provides another telling example of the significance of process innovation: a case, argue Kline and Rosenberg, in which "the subsequent improvements in an invention after its first introduction may be vastly more important [economically]."[21] As Joel Mokyr describes, in the nineteenth century steel was prohibitively expensive, until Henry Bessemer invented a process that removed impurities from steel, enabling low-skilled workers to produce steel in much greater quantities.[22] Bessemer steel once again improved dramatically in quality after Robert Mushet added tungsten and manganese to it. The Siemens-Martin process yielded further improvements, as did the efforts of Percy Gilchrist and Sidney Thomas—two British tinkerers who added limestone, without which "the German steel industry could never have developed as it did." Furthermore, "American inventors added a number of other improvements, such as 'hard driving' . . . improved blowing engines; and direct casting of the pit iron into the steelworks."[23] The industry's development reflects decades of incremental innovations that led to dramatic increases in output.

The steel industry example also shows the multiple levels at which innovations occur. Amar Bhidé explains that innovation occurs in high-level scientific principles, mid-level technologies, and ground-level "context-specific heuristics or rules of thumb."[24] Innovation at any of these levels connects with (and indeed, requires) innovations at others. "A breakthrough in solid-state physics has value in the semiconductor industry only to the degree that it is accompanied by the development of new microprocessor designs; and the new designs may be useless without the development of plant-level tweaks for large-scale production of the microprocessor."[25]

Keith Pavitt created a taxonomy of innovation that occurs in different areas of an economy. For example, many people think of innovation as happening at government labs at which white-coated geniuses squint into microscopes and pour liquid into test tubes. Indeed, science-based innovation (Bhidé's "high-level" innovation) connects with commercial innovation. As Pavitt argues, basic science laid the foundation for product innovation: "synthetic chemistry and biochemistry for the chemical industry . . . electromagnetism, radio waves and solid-state physics for the electrical/electronic industry."[26] Today, China's dominance in the electric vehicle battery industry stems from its scientific advancements in chemistry and metallurgy.[27] Science-based innovation emphasizes basic research and certain sectors such as biotechnology, chemicals, semiconductors, and pharmaceuticals.[28]

Other types of innovation (Bhidé's mid- or ground-level innovation) are more applied, and are performed by firms seeking new products, or searching for ways to shave time and money off their production and delivery processes. Pavitt describes "specialized suppliers" or "supplier dominated" innovation.[29] Researchers at the McKinsey Global Institute characterize the "design and engineering of new products" as engineering-based innovation.[30] Such activities are generally performed by private firms, for example, in commercial aviation, auto manufacturing, and telecommunications hardware. For example, an aerospace firm might seek to improve an airplane's fuel consumption by engineering a new kind of wing that creates less wind resistance. Firms also engage in efficiency-driven innovation, seeking to improve existing products: raising product quality, reducing production costs, and reducing time to market.

Scholars also consider demand-based or customer-focused innovation, in which entrepreneurs observing market demand improve existing products or adapt them to new markets.[31] This kind of innovation is concentrated in industries such as internet software and services, appliances, and consumer goods. Innovators may be well-known firms, or entrepreneurs in their proverbial garages. In sum, innovation occurs at all levels of an economy, and innovation at one level supports innovation at others.

Measuring Innovation: Inputs and Outputs

Comparing countries' innovation performance requires a method for measuring innovation, which is a daunting challenge. Any given metric might capture some innovative activities while neglecting others.[32] The best approach is through triangulation—the use of multiple metrics that capture different dimensions of innovation. Scholars assessing national innovation often look to two broad areas: metrics related to inputs (i.e., the efforts a country is making to encourage innovation) and metrics related to outputs (the extent to which this effort is yielding results).[33]

Several metrics are proxies for the extent to which a government is creating conditions (described in Chapter 1) conducive to innovation. To capture government investment in S&T capabilities, scholars rely on the metric of government spending on research and development (R&D). Given the importance of tertiary education and universities in innovation—for invention as well as for diffusion[34]—other metrics include a country's education spending as a fraction of gross domestic product (GDP), and the quality of universities as reflected in global university rankings.[35]

Because innovation depends on a well-trained S&T workforce, other important metrics seek to capture this key input. "PhD-level experts make up a small but important component of the STEM workforce, spearheading research and development efforts that push the boundaries of their fields and educating the next generation of science and technology leaders."[36] I thus assess a country's number of S&T PhDs, as well as the number of S&T bachelor's degrees awarded.

While input metrics reflect the extent to which a country is trying to develop cutting-edge technology, output metrics capture the extent to which it is actually succeeding. First, analysts use patent data to capture whether a new technology represents a meaningful advancement. One of the most frequently used sources is the World Intellectual Property Organization (WIPO), which compiles data from national patent offices. In the United States, the National Science Board (NSB) relies on this indicator to track national innovation levels.

Patents are a tricky indicator; national patent offices vary in the quality of inventions for which they grant patents.[37] In 2024 China filed more patents than any other country, with over 47 percent of the global total.[38] Scholars note that because Chinese government subsidies encourage firms to apply for patents for negligible technological improvements, Chinese national patent data—represented in WIPO data—probably exaggerates Chinese innovation.[39]

Thus scholars prefer to measure innovation via patents that are granted not only domestically but also in multiple jurisdictions. A harder test of Chinese innovation would be patents granted to Chinese firms by the US Patent and Trademark Office (USPTO). An even harder test would be "triadic" patents: those filed jointly with the USPTO, the Japan Patent Office, and the European Patent Office. An invention that is granted patents in all three offices is highly likely to be truly innovative.[40] Such patents probably understate Chinese innovation activity, however. Because they are more expensive and have longer lead times for approval, they tend to be sought only by large, well-established firms (thus failing to capture significant innovation by small- and medium-sized enterprises). In sum, these three metrics range from an easier test (the WIPO) to a harder test (USPTO) to a much harder test (triadic patents).

Second, scholars measuring innovation also assess the extent to which a country is producing high-quality research. Counting journal articles risks counting articles that have limited scientific value; as with patenting, some governments subsidize (and universities promote) faculty on the basis of publication quantity, which has led to the creation of many low-quality or even fake journals. To measure the extent to which research is innovative, scholars thus rely on a bibliographic metric that counts how many S&T articles a country's experts publish in scholarly journals. As a marker of quality, this number is weighted by how many citations those articles receive. Articles that are more frequently cited are viewed as having a larger impact on the field.[41]

Third, other metrics seek to reflect the extent to which innovative technology is successfully diffusing through the economy and being commercialized by firms. Scholars note that one of the most important proxies for diffusion is a country's S&T workforce—an input metric described above.[42] Furthermore, I assess two metrics related to commercialization: a country's

Table 2.1. Indicators of national S&T capabilities

Concept	Metric	Source
Size and quality of S&T workforce	Number of university degrees in S&T	National Science Board
	Number of doctoral degrees in S&T	
Participation in global knowledge networks	Extent of international collaboration/coauthorship	National Science Board
	Coauthorship with US scholars	
Level of investment in education	Education spending as percentage of GDP	World Bank
	Tertiary education spending	
High-quality universities	Global ranking of universities	US News & World Report, THE rankings
Government's effort to encourage innovation	R&D as a percentage of GDP	OECD, National Science Board
	Aggregate R&D spending, country comparison	
Research that rises to the level of being a new and influential technology or idea	Triadic patents	OECD, National Science Board
	USPTO patents	
	International patents (WIPO)	
	Number of S&T publications weighted by citations	
Success in diffusing and commercializing innovation	Value-added exports as a percentage of total exports	World Bank, National Science Board
	Share of global knowledge and technology intensive (KTI) exports	

percentage of value-added exports, and "high-technology"/"knowledge-intensive" exports.

In sum, scholars measure the elusive concept of innovation by triangulating with multiple metrics. Because each metric captures a slightly different part of the picture, together they do a good job representing national innovation. Table 2.1 summarizes these metrics.

The CCP's Innovation Push

The CCP devotes significant attention, rhetoric, and resources to encouraging innovation.[43] In Five-Year Plans, other government initiatives, and leaders' and party statements, the CCP describes S&T capabilities as essential to encourage economic growth and to compete in a dangerous geopolitical world.[44] Early in China's economic rise, Deng Xiaoping, Zhao Ziyang, and other Chinese leaders studied and were galvanized by what they called the

"New Industrial Revolution." As Julian Gewirtz describes, China's ability to compete with the United States, Europe, and other hi-tech leaders was deemed "a matter of life and death of the nation."[45] In 1986, Chinese scientists proposed, and the Chinese government adopted, the "863 Program," which identified a range of key technological areas in which China needed to improve its capabilities.[46]

As China's economy grew, the CCP continued its technological push. The CCP noted in a 2006 initiative that other countries "have made S&T innovation a national strategy and S&T inputs strategic investments by drastically increasing R&D spending. These nations lead the world in deploying and developing frontier technologies and strategic industries and implement important S&T programs in an attempt to enhance their national innovative capability and international competitiveness." In this world, China would need to cultivate innovation as "a major driving force" for the country's economic and social progress.[47] In a 2014 speech to prominent scientists and scholars, leader Xi Jinping argued that achieving "the great rejuvenation of the Chinese nation" required "science and education and innovation-driven development." He argued that "the competition for comprehensive national strength" would be decided by science and technology.[48]

Another important policy was the "Made in China 2025" plan. Announced in 2015, this plan represented China's push to become a "science and technology innovation superpower." The plan identified and subsidized, with billions of dollars, ten specialized fields.[49] These include next-generation IT, high-end numerical control machinery and robotics, aerospace and aviation equipment, maritime engineering equipment and high-tech maritime vessel manufacturing, advanced rail equipment, energy saving vehicles, electrical equipment, agricultural machinery and equipment, new materials, and biopharmaceutical and high-performance medical devices. To encourage the development of these technologies, the CCP provides research subsidies, government procurement, and low-interest loans from government-controlled banks and investment funds; it supports foreign acquisitions and the recruitment of foreign experts.[50] The CCP has also created 522 "national key laboratories" as well as 350 "national engineering research centers."[51] Criticism from overseas encouraged Beijing to play down the "Made in China 2025" name, but the initiative continued in its goal of making China a technology superpower.

China's innovation push came together in what is called the "Innovation Driven Development Strategy," articulated in 2016. Tai Ming Cheung describes this strategy as resting on five pillars: (1) market-driven innovation that would be supported and coordinated by the state; (2) an emphasis on indigenous innovation; (3) cultivating the highest levels of human capital, and striving to attract high-skilled workers from overseas; (4) creating an innovation-boosting regulatory environment (in terms of intellectual property protections, capital markets, and taxation), and finally, (5) expanding international cooperation.[52]

Xi underscores his government's commitment to innovation in high-profile speeches. At the 19th National People's Party Congress in 2017, Xi pledged to make China a global innovation leader by 2035. "We should aim for the frontiers of science and technology, strengthen basic research, and make major breakthroughs in pioneering basic research and groundbreaking and original innovations."[53] At the 20th Party Congress, Xi stated, "We will accelerate the implementation of the innovation-driven development strategy, will speed up efforts to achieve greater self-reliance, and strengthen science and technology to meet China's strategic needs."[54] Careful observers noted that Xi's speech, as well as other policy documents, emphasize "technology" rather than "growth," reflecting a priority shift.[55] After the United States imposed export controls on semiconductors and related equipment, Xi described the hi-tech sector as the "main battlefield" of superpower competition; he declared that to reduce China's vulnerability, it must strive to excel in specific areas such as industrial machine tools, advanced materials, and integrated circuits.[56]

The CCP also promotes Chinese technology through leadership in global standards-setting.[57] Domestically, with any new technology, rapid and centralized standards-setting allows industry to develop more efficiently, and to benefit more quickly from economies of scale. Globally, a country's leadership in standards-setting promotes the market dominance of its firms.[58] For example, developing countries seeking to catch up "have had to work under the umbrella of Western rules and regulations governing major technologies."[59] With this in mind, Beijing is working hard to set global standards. In international bodies such as the influential International Standards Organization (ISO) and International Electrotechnical Commission (IEC), Chinese nationals increasingly occupy secretariats of technical committees. "Those in a leadership capacity are able to influence the agenda, how conversations are structured, and how time is allocated."[60] Furthermore, Chinese aid and trade within the Belt and Road Initiative encourage future demand for Chinese technology configured to those standards. Overall, standards-setting is an important "weapon to eventually displace the West in the high-stakes technology battle."[61] In sum, through both internal policies and diplomatic activity, the CCP has shown a strong commitment to developing world-class S&T.

CHINA'S INNOVATION INPUTS

Data on innovation inputs reflect the CCP's innovation push. First, China's R&D spending has increased over time. The world's highly innovative economies spend about 3 percent of GDP on research and development. China began increasing its R&D spending in the late 1990s, and as a percentage of GDP it rose from 0.83 percent in 2000 to over 2 percent by 2014 (surpassing the European Union average).[62] The Chinese government plans for R&D spending to rise to 2.8 percent of GDP by 2030.[63]

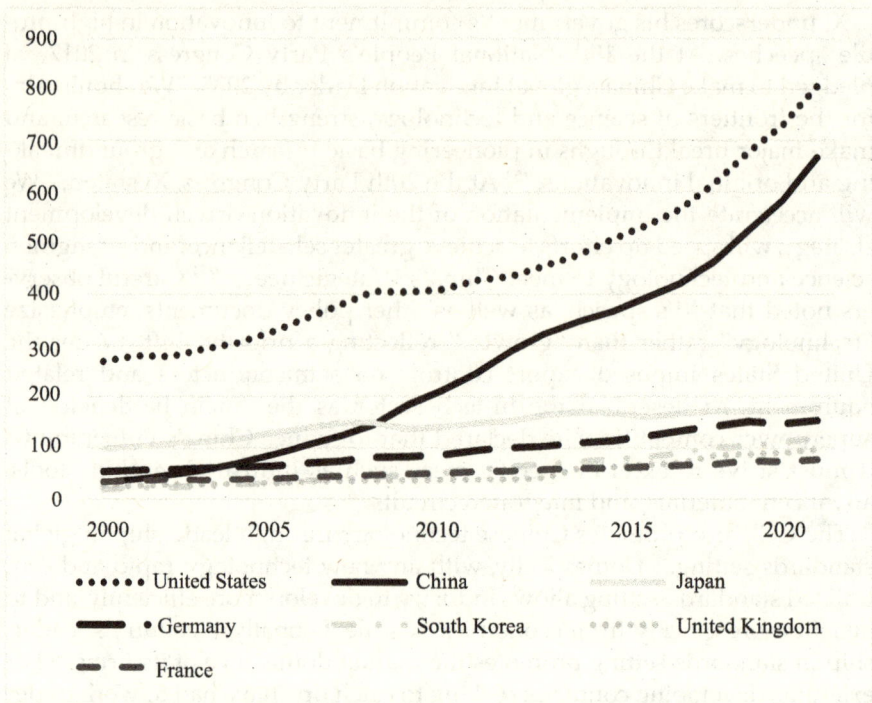

Fig. 2.1. Aggregate R&D spending by country.
Source: National Science Board.

Rush Doshi, a scholar and former Biden administration official, argues that China has learned from the United States how to encourage S&T innovation. "Beijing recognizes, as the United States once did, that such research cannot be supported entirely by the market and the private sector and instead must be supported by the public." The CCP has poured resources into R&D as a result, Doshi observes, particularly in sectors relevant to the fourth Industrial Revolution, where "China spends roughly $2.5 billion annually, a modest sum that is nonetheless estimated to be more than ten times what the U.S. spends" in similar areas.[64] Figure A in the online appendix shows the rise of Chinese R&D spending as a percentage of GDP over time. Today, as measured by this indicator, China has risen from being a low spender on R&D to topping the amount spent by the European Union average.

Second, China's massive GDP means that in aggregate terms, China's R&D spending (at $668 billion) is second only to that of the United States ($806 billion).[65] See figure 2.1.

Third, the CCP has presided over a dramatic transformation in Chinese tertiary education. Mao Zedong's Cultural Revolution—which targeted "anti-cosmopolitan" forces and privileged ideology over any logical or practical approach—eviscerated China's already weak universities. As William Kirby writes, "From 1966 to 1969 enrollment of new students was stopped completely, intellectuals were persecuted, and several universities became political and military battlegrounds"; after 1970, when admissions were reinstated, students were admitted on the basis of "redness" (political correctness) rather than their scores on entrance examinations.[66]

Education reforms since the Deng Xiaoping era have dramatically improved the quality and accessibility of Chinese tertiary education. Kirby notes that acreage dedicated to universities in China has quintupled since 2000, and that universities have spread across the country. "Until recently, higher education institutions had been concentrated in but a few areas, Beijing and the lower Yangzi region chief among them," Kirby notes. "Today cities and provinces never known for higher education compete to found, build, and expand colleges and universities—often within new science and technology zones—as part of their competitive strategies for growth, development, and prestige."[67]

Reform of tertiary education has included spending increases. Education spending rose from 1.9 percent of GDP in 1980 to 4 percent in 2022.[68] This puts China still below the Organization for Economic Cooperation and Development (OECD) average (5 percent), and among other prominent innovators such as France, Germany, Japan, South Korea, and the United States—which spend 4~5 percent of GDP on education.

Fourth, the CCP is also committed to developing an S&T workforce. University attendance in general is skyrocketing; from 2000 to 2023, participation rates in tertiary education rose from 8 percent to 75 percent.[69] Chinese universities now graduate more than 12 million students per year.[70] Since 1990, China's university-educated workforce "has exploded": China's college-educated workforce will likely overtake that of the United States by 2025, and could be double its size by 2040.[71] The number of bachelor's degrees awarded in S&T fields is also growing. Figure B in the online appendix shows China's undergraduate S&T degrees over time.

China is also graduating growing numbers of S&T PhDs. In his remarks to the 19th Party Congress in 2017, Xi Jinping said that China must become a "nation of innovators" in order to foster economic growth: that China needed to train and recruit scientists in strategically significant fields. A report on China's STEM workforce notes that the Ministry of Education "roughly doubled its spending on higher education between 2012 and 2021, fueling an increase in new PhD enrollments"; across the country the government funded "more than 1,300 new PhD programs at dozens of institutions that had previously not offered doctoral programs." The report concludes, "China's capacity to produce skilled PhD-level STEM experts appears to be growing rapidly."[72] Figure C in the online appendix shows the increase in China's science and engineering doctoral degrees, which has led it to overtake the United States in this category.

Part of the CCP's push to develop high-quality human capital involves countering brain drain and attracting foreign talent. Hundreds of thousands of Chinese students receive their training overseas, and the CCP created the "Thousand Talents" program to encourage them to return home to raise China's level of indigenous S&T.[73] Analysts report problems with the program, such as corruption and lackluster results.[74] But Andrew Kennedy notes that the challenge of brain drain has evolved over time. For China, India, and other developing countries, he observes, "flows of human capital are no longer simply a unidirectional brain drain": "well-educated migrants now frequently visit or return to their home countries, and in some cases the brain drain turns into 'brain circulation.'"[75] China has also had success attracting foreign talent; Chinese students are educated by many professors "who received their education and training in the United States, United Kingdom, Germany, Singapore, and Japan."[76] Furthermore, in response to US-led semiconductor export controls, China has aggressively recruited foreign S&T workers in the semiconductor sector both by bringing them to China and by setting up R&D centers in South Korea, Taiwan, the United States, and other countries.[77]

Fifth, the CCP has also sought to raise the quality of Chinese universities. Upon the founding of the People's Republic of China (PRC), Chinese universities were cut off from the West and modeled on the Soviet system. For many years Chinese universities lagged elite universities in Japan, Singapore, South Korea, Europe, and the United States. Today a quality gap remains, but the CCP is making determined efforts to improve, with signs of success.

China's universities are climbing in rankings such as Times Higher Education (THE), a survey of ten thousand top scholars in the world. "Where once the list of universities with the highest scientific impact would have been dominated by U.S. and U.K. schools including Cambridge, Stanford, Harvard and MIT, the new top 10 list of universities with high scientific impact includes six universities from China."[78] According to THE's 2024 overall rankings, Tsinghua University outranked the University of Chicago; Tsinghua and Peking University outranked Columbia, Cornell, UCLA, and the University of Pennsylvania.[79] As for universities within emerging markets, THE ranked Chinese universities in seven of the top ten spots (with Tsinghua and Fudan in the top two positions), and 81 Chinese universities appear on the list (the most of any country).[80] Under a 2015 program called "World Class 2.0," the CCP seeks to further improve China's top 9 universities, aiming to have all 9 ranked among the world's top 15 universities by 2030.

Finally, China has developed massive and highly innovative domestic clusters, and enjoys strong connections to global business and knowledge networks. Chinese cities now rank in the top ten of global clusters. Three Chinese cities have the world's highest amount of venture capital investment (Beijing, Shanghai, and Shenzhen), three have the highest number of unicorn startup companies (Beijing, Shanghai, and Hangzhou), and two Chinese areas lead in patent filings (Shenzhen-Hong Kong and Beijing).[81] The Pearl River Delta region in southern China, with nearly 60 million people, "resembles Silicon Valley in being not a singular geography but a globally interconnected one, attracting foreign investment and collaboration and exporting its influence on the world's technology infrastructure."[82]

Chinese students and researchers also have a high degree of interaction with cutting-edge researchers around the world. China sends the most students abroad of any country, with over 700,000 students studying overseas each year. And Chinese researchers often collaborate with overseas partners. The State Key Laboratory System—China's state-funded network of nearly 500 laboratories—is "one of the most important building blocks in China's innovation base," connecting private firms, state-owned enterprises (SOEs), and universities. The labs foster exchanges, visiting scholar programs, conferences, and coauthoring between Chinese and overseas scientists.[83]

The private sector in China also enjoys significant interaction with global knowledge communities. Chinese firms have opened up several research laboratories around the world, which enables them to recruit the world's top personnel. Tencent operates research labs in Seattle; DiDi Chuxing has one in Mountain View, California; Baidu has two labs in Silicon Valley; and Alibaba has spent billions to set up research labs in Singapore, Moscow, Tel Aviv, Seattle, and Silicon Valley.[84] Likewise, foreign firms have set up research institutes in China. In 2024, the *Economist* called China "the world's research-and-development laboratory," with foreign firms doubling their R&D spending in China since 2012. "Foreign chief executives now believe," the newspaper commented, "that China's brainpower and its innovation-curious regulatory regime are crucial ingredients of their companies' global success."[85]

China's extensive interaction with researchers at the global cutting edge creates opportunities for coauthored research. Figures D and E in the online appendix show the extent of Chinese coauthorship; the vast majority of the coauthors have been US citizens.[86]

In sum, input metrics show strong Chinese S&T performance across multiple dimensions.

The Sun and the Moon

China made a sun. In 2022, its EAST (Experimental Advanced Supercomputing Tokamak) nuclear fusion reactor maintained a temperature of 126 million degrees Fahrenheit (about five times as hot as the sun) for about 17 minutes, obliterating previous records set by other scientists around the world. In

2025, the EAST reactor smashed its own record.[87] For decades scientists have been trying to harness the clean energy released by nuclear fusion—that is, what powers the stars. China has moved the world a step closer to what Stephen Hawking described as an "inexhaustible supply of energy, without pollution or global warming." Chinese scientists are joining those from 35 other countries in building the ITER (International Thermonuclear Experimental Reactor) in France. This reactor will be the world's largest.

China has numerous other globally prominent scientific achievements. Its space program has "steadily hit milestones": Its Chang'e missions landed on the far side of the moon, collected soil samples with a rover, and "in a first for any country" successfully returned those samples from the far side of the moon to earth.[88] The *Washington Post* notes that "Beijing has ambitions to become a space power and a scientific force, laying out plans to land Chinese astronauts on the lunar surface by 2030 and set up a base at the moon's south pole."[89] China's investments in supercomputing have paid off to the point that it is battling the United States for dominance, leading to headlines such as "US Rushes to Catch Up to China in Supercomputer Race."[90] In the field of quantum communications, China has taken a—*what's another way to say a "great leap forward"?*—by establishing in 2017 "the world's first quantum-encrypted intercontinental video link.[91]

In addition to these scientific achievements, Chinese firms have become leaders in high-tech products. China excels in telecommunications (with Huawei globally dominant), high-speed rail, and LCD panels. Chinese firms now dwarf all other competitors in the global solar panel industry, with dramatic cost advantages. Chinese leaders say that a "'new trio' of industries—solar panels, electric cars and lithium batteries—has replaced an 'old trio' of clothing, furniture and appliances."[92]

Indeed, Chinese firms now dominate the global electric vehicle (EV) market, and China is the world's leading producer of EV batteries.[93] A common refrain holds that China is shipping copycat products overseas at prices slashed by government subsidies. Chinese firms do indeed benefit from subsidies—but the competitiveness of Chinese EV batteries is also a reflection of significant innovation in both power and design. Explains Xu Zeyu, the Chinese manufacturing company BYD's dramatic innovation of "blade" batteries raised the energy density of power cells, sending "shock waves across the EV battery market."[94] Two years after this innovation, "BYD dethroned Tesla as the world's biggest EV manufacturer." More powerful batteries have extended vehicle ranges and permitted heavier cars, leading Chinese designers to create models with greater comfort and more options.[95] As the Semafor news website concludes, "The world is lagging behind China on EV battery innovation to the extent that global manufactures are barely able to compete."[96] In part because of its EV dominance, in 2023 China shocked the world by overtaking Japan and becoming the world's top automobile exporter.[97]

China performs particularly well in customer-based innovation, in which it benefits from its massive, geographically diverse market and "hundreds of millions of hyper-adoptive and hyper-adaptive consumers."[98] Haier is now a leading global brand and the largest global producer of consumer appliances; *Fast Company* named it one of the world's most innovative firms, and McKinsey called it "a pioneer in both organizational and customer-centric innovation."[99] China adapted military aerial vehicle technology to create—and dominate—the commercial drone market. Chinese e-commerce firms are using drones to extend delivery ranges and reduce delivery times.[100] Furthermore, Chinese firms operate at the cutting edge in financial technology.[101] The highly competitive sector used digital technology to create "inclusive" financial products and services, connecting to the economy hundreds of millions of Chinese people who previously had no financial history. China is the global leader in mobile payments; digital payments have taken over Chinese society, with dominant apps—WeChat Pay and Alipay—logging a billion or more users each. Chinese firms have also adopted the "big-data-supported credit risk assessment model" that relies on AI.[102] China has eight fintech "unicorns" with a market valuation of over $200 billion. The country's fintech boom ended when the CCP stepped in to provide greater oversight; however, government restrictions have since eased and China remains a global leader in this sector.[103]

CHINA'S OUTPUT METRICS

China's S&T achievements are captured in its impressive performance across global innovation metrics. First, China is experiencing a surge in patent activity, not only in domestic patents but also in international patent offices. WIPO reports that China is the number one country in terms of patenting applications (with 46.6 percent of the global total; the United States is second at 17.4 percent). The rest of the top five are Japan, South Korea, and the European Patent Office. China's Huawei is the global leader in patent filings, with South Korea's Samsung a distant second.[104] According to WIPO, China is the number one country in terms of patents granted.

As described earlier, WIPO patent data does a good job capturing the broadest picture of Chinese innovation, but likely exaggerates it by including lower-quality patents.[105] Turning to other patent data, US Patent Office data show that China has risen significantly in its number of patents granted over time, such that China is now first in the world at nearly a 50 percent share.

Another patenting metric—triadic patents, the highest bar of the three—similarly shows that China has emerged as an innovation leader. China now receives the world's fourth-largest number of triadic patents (see figure F in the online appendix). On this metric China remains far short of Japan and the United States—as do the rest of the pack. However, China

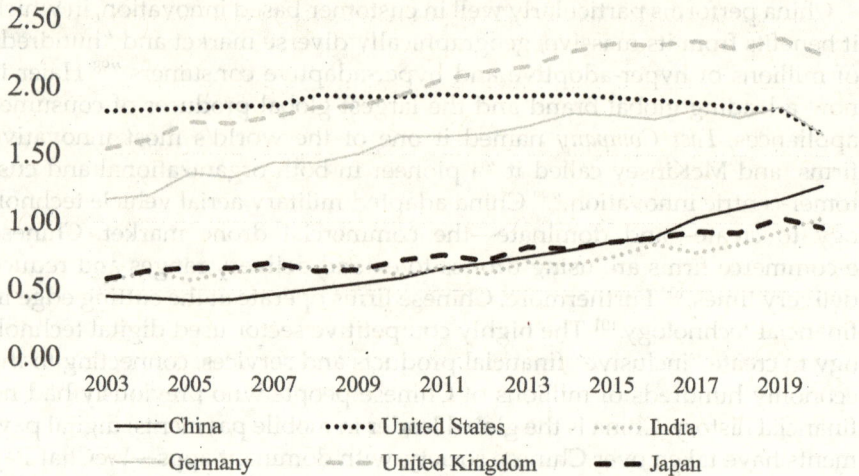

Fig. 2.2. S&T publications in top 1% of cited publications, 2003–2020.
Source: National Center for Science and Engineering Statistics, National Science Foundation, 2024.

has overtaken highly innovative countries such as France, Germany, South Korea, and the United Kingdom.

As described earlier, triadic patents capture patents across multiple jurisdictions. Given home-country bias, China's competitiveness in the USPTO, the Japan Patent Office, and the European Patent Office is a significant indicator of quality. Importantly, triadic patenting (with higher costs and longer lead times) probably understates Chinese innovation. In sum, the truth is somewhere between these three metrics, all of which show that China has emerged to be a leading innovator.

Observers have been startled by reports of China's rapid rise in scientific research.[106] Indeed, the metric of highly cited S&T research publications reflects significant Chinese improvement over time.[107] The index in figure 2.2 measures a country's number of highly cited S&T articles relative to the world's highly cited (i.e., top 1 percent) of articles. The figure shows the world average as 1.0. According to this metric, China has significantly improved its performance in the past two decades, exceeding the world average and now authoring the third-largest number of highly cited articles, below the United States and the European Union.

The Nature Index—a series of rankings of countries' technological excellence—shows that after years of rising through the rankings, China in 2023 edged out the United States to occupy the overall top position.[108] In the Nature Index ranking of research institutions (ranked by their article output in the natural and health sciences), Chinese institutions have risen rapidly—to the point that in 2024 China had seven institutions among the top ten, with the Chinese Academy of Sciences occupying the top spot, edging out Harvard University. The US, French, and German rankings in the top ten have all declined.[109] The Nature Index described China's rise in the rankings for high-quality science publications as "meteoric."[110] Furthermore, along with the United States, China is a world leader in AI research. "It is clear," testified one analyst, "that China has become a leader in AI research, based on metrics such as the number of research publications, citation counts, and participation in top AI conferences."[111]

As noted, the extent to which a country can diffuse and commercialize innovation is a key dimension of its technological capabilities.[112] Human capital is a key driver of diffusion; as discussed earlier, China has a massive talent pool of scientists and engineers for successful diffusion. China also leads the world in AI engineers, which suggests that it is well-situated to diffuse this key emerging technology.[113] Furthermore, China is successfully commercializing innovative technologies, as seen in the metric of high-value-added manufactured exports (as a share of total exports). As Figure G in the online appendix shows, on this metric China performs above the OECD average, and exceeds the performance of most countries.

A related metric examines China's share of world "knowledge and technology–intensive" (KTI) exports. Figure H in the online appendix shows that China has been increasing its global share of high technology exports over the past two decades.

Some scholars argue that high-tech exports are a poor indicator of indigenous S&T capabilities, particularly in China's case. They argue that these reflect not Chinese S&T capabilities but rather the capabilities of the Taiwanese, German, or US firms that export from China. Importantly, however, the share of China's high-tech exports produced by foreign firms has fallen dramatically. A study done by the Center for Strategic and International Studies (CSIS) shows that the foreign share of Chinese high-tech exports fell from 70 percent in 2011 to 25 percent in 2022.[114] Furthermore, Taylor argues that the quality of labor at Chinese factories is high, and he notes that foreign firms locate in China in part because of the technical capabilities of the local workforce.[115]

In sum, China has developed the S&T capabilities to close the gap with the world's most innovative economies. China's government has been committed to encouraging S&T, has provided necessary innovation inputs, and has experienced remarkable success as measured by a variety of innovation metrics. The country once dismissed as a copycat has become a serious S&T player.

The Fourth Industrial Revolution

Can China sustain its cutting-edge S&T performance as the world moves into a new technological era? The advent of AI and other emerging technologies has led many observers to argue that the world is entering the "Fourth Industrial Revolution," or 4IR. "A global technology revolution is now underway," US former Secretary of State Anthony J. Blinken said. "The world's leading powers are racing to develop and deploy new technologies like artificial intelligence and quantum computing that could shape everything about our lives—from where we get energy, to how we do our jobs, to how wars are fought."[116] Will China's technological competitiveness continue—or is China likely to fall behind as productivity gains and military advantage increasingly come from a new set of technologies?

Although the technologies that people associate with the 4IR vary, AI is generally seen as one of the most transformative: a "general-purpose technology" that has the potential to transform every sector of the economy.[117] Computers using AI perform functions associated with the human brain: They can think, recognize, learn, infer patterns, and make predictions. They do so by churning though massive data sets, learning as they go, and then applying what they learned. The field of AI encompasses many different functional areas, including facial recognition, autonomous vehicles, and voice recognition and natural language processing (NLP).

At this early stage, various metrics show the United States leading in AI, with China a serious rival. "The United States of America is the leader in AI, and our administration plans to keep it that way," declared Vice President J.D. Vance in a 2025 speech in Paris. "The U.S. possesses all components across the full AI stack, including advanced semiconductor design, frontier algorithms, and, of course, transformational applications."[118] Vance's confidence was not misplaced; Stanford's AI Index found that in 2024 "the U.S. has the world's most robust AI ecosystem and outperforms every other country by significant margins," with China in second place.[119]

The United States, according to the Stanford index, leads the world in high-quality AI research published, dominance of machine learning models, private investment in AI, and the most merger and acquisition activity.[120] Furthermore, the United States and its security allies continue to dominate the technologies—semiconductors and semiconductor manufacturing equipment—vital to training AI models, although China has made impressive gains since the United States began imposing export controls.[121]

Though the United States remains in the lead, numerous indicators show China to be the United States' closest AI competitor. "China's AI boom," proclaims the *Economist*, "is reaching astonishing proportions."[122] Chinese firms now routinely grab headlines and startle markets with notable innovations. For example, DeepSeek's 2025 AI model, trained at a fraction of the cost of comparable models, roiled US stock markets, and Butterfly Effect's Manus model unleashed "more hype than a Taylor Swift concert."[123] Chinese and US AI firms are roughly equal in total value (with China at $160 billion with 13 major firms, and the United States at $176 billion with 33 firms).[124] China is also leading the world in AI-related patents.[125] Brummer and Lind show that China has emerged as the second-largest holder of high-quality patents related to 4IR technologies, and is second in producing high-impact publications on AI-related research.[126] Furthermore, the Chinese Academy of Sciences is now the global leader in AI research output (including the most-cited research).[127]

In sum, not only is China a current S&T leader but also its AI capabilities suggest that its leadership will continue into a new technological era. At this early stage, one must make predictions with tremendous humility given uncertainty about which 4IR technologies will emerge as the most significant—let alone which countries will dominate them. Furthermore, many observers argue that China's future S&T performance will suffer because of policy changes under Xi Jinping (a topic I explore in Chapter 6).[128] However, at the present time it's clear that China is competing hard, to great success.

Military Technology

China's development of cutting-edge military technologies also supports the claim that the country is a capable great-power competitor. Over the past few decades, Chinese military modernization has increasingly threatened the regional dominance of the US military.[129]

The People's Liberation Army Air Force (PLAAF) has rapidly transitioned from flying antiquated MiG-21s to building and flying fifth-generation stealth J-20s. The PLAAF has acquired over 800 fourth-generation fighters (the J-10, J-11, and J-16) as well as a long-range, standoff, precision bombing capability: bomber/attack aircraft that can strike Guam. China is improving its reconnaissance capabilities, with early-warning aircraft and with a growing fleet of unmanned aerial vehicles (UAVs). China's long-range strike and intelligence, search, and reconnaissance (ISR) capabilities increasingly challenge the US military's ability to move its vessels around the region safely and to defend Taiwan in the event of a Chinese attack.[130]

China has also modernized and grown its naval capabilities. The People's Liberation Army Navy (PLAN) is now the world's largest navy in terms of tonnage, commissioning four aircraft carriers. The sophistication of these carriers (i.e., the electromagnetic catapult–assisted launch system) shows China catching up to the United States' lead in carrier aviation. China's Type 055 destroyer (a *cruiser*, according to the NATO definition) is also highly advanced, described by a RAND analyst as having "a sophisticated design, stealth features, radars, and a large missile inventory."[131]

China has also developed a range of technologies that increase the lethality of its conventional forces. A major component of Chinese military modernization has been "informatization": Oriana Skylar Mastro writes that toward that goal, China has developed capabilities in "space command, control, communications, computers, intelligence, surveillance, and reconnaissance (C4ISR) infrastructure to support its informatized needs, from navigation and communication to precision guidance and real-time command and control" as well as "modern conventional platforms with which to conduct network-centric warfare."[132] The People's Liberation Army's (PLA's) C4ISR system relies on the largest global constellation of satellites controlled by any country other than the United States. To fight informatized warfare and to carry out long-range precision strikes, China in 2015 established the "Strategic Support Force," which oversees "kinetic, cyberspace, space, electromagnetic, and psychological operations."[133] China's cutting-edge missile systems, such as the DF-17 hypersonic missile and the DF-26 "carrier killer" air-launched, intermediate-range ballistic missile, have significantly elevated the regional threat to the US Navy.

Furthermore, China has transformed itself from a third-rate nuclear power—which could barely claim an assured retaliation capability—to one of the world's three leading nuclear states.[134] China's nuclear force is growing faster than any other arsenal in the world. Its quantitative growth is matched by technological breakthroughs in delivery systems—potentially including hypersonic missiles and fractional orbital bombardment systems—that would give China meaningful advantages in its nuclear competition with the United States.[135]

In sum, decades of military modernization have transformed China from a technologically backward military emphasizing ground power to a sophisticated maritime military power. These capabilities are threatening the access that the US military has long enjoyed in the region. In some areas, China is even pulling ahead of the United States. A study by the Australian Strategic Policy Institute (ASPI) noted that China "has a commanding lead in hypersonics, electronic warfare and in key undersea capabilities." ASPI further warned that "China's leads are so emphatic they create a significant risk that China might dominate future technological breakthroughs in these areas."[136]

4IR MILITARY TECHNOLOGIES

Emerging technologies offer significant potential for military applications. Among them, both the US military and the Chinese military are investing heavily in AI. "A growing number of robotic vehicles and autonomous weapons are able to operate in combat zones too hazardous for human combatants," comments a team from RAND. "Intelligent defensive systems are increasingly able to detect, analyze, and respond to attacks faster and more effectively than human operators can. And big data analysis and decision support systems offer the promise of digesting volumes of information that no group of human analysts, however large, could consume and thereby help military decisionmakers choose better courses of action more quickly."[137] Furthermore, as Elsa Kania writes, "Advances in autonomy and AI-enabled weapons systems promise to increase the speed, reach, precision, and lethality of future operations." Unmanned aerial vehicles can potentially operate autonomously rather than via remote control; AI-equipped missiles could become more autonomous in order to improve their accuracy.[138] Furthermore, as RAND comments, "Autonomous systems will better enable friendly forces to operate in A2/AD [anti-access/ aerial denial] environments. Not only will they reduce the numbers of human operators at risk in these environments, but they can also be made smaller, faster, and more agile than inhabited weapons platforms and thus potentially more combat capable." Autonomous weapons and intelligence, search, and reconaissance (ISR) platforms, RAND summarizes, "will be able to operate in areas that humans cannot."[139] The US military is developing AI applications for long-range antiship missiles (which can autonomously select and engage targets), autonomous flying vehicles (the "Loyal Wingman" program), as well as for logistics and planning and numerous other areas. The Defense Advanced Research Projects Agency (DARPA) has offered prizes as a means of encouraging AI-related research and has established a "Defense Innovation Unit" in Silicon Valley to encourage industry cooperation with military research.[140]

The Chinese government had declared that "informatized" warfare is shifting to "intelligentised" warfare, and they see AI as integral to China's future military success. Xi Jinping "personally emphasized the emergence of this 'new RMA.' . . . He called for China to advance military innovation in order to 'narrow the gap and achieve a new leapfrogging as quickly as possible.'"[141] In 2017, AI was identified as a core strategic technology in China's New Generation Artificial Intelligence Development Plan. Since then, notes Kania,

> this agenda has progressed at all levels of government and through the efforts of a range of stakeholders. China's AI efforts have built upon and harnessed the robust efforts of China's dynamic technology companies. This plan also discussed the implementation of a strategy of military-civil fusion in AI, calling for strengthening its use in military applications that include command decision-making, military deductions (e.g., wargaming), and defence equipment.[142]

Across all of its branches, the PLA has been introducing robotic and unmanned vehicles with varying levels of autonomy; the PLA Rocket Force and newly centralized Strategic Support Force that has combined cyber, electronic, psychological, and space warfare are similarly relying on greater autonomy.[143]

China has a significant military drone program (aided by its status as the number-one commercial drone producer). "China has worked to enable its drones to operate with greater autonomy with capabilities for taking off, landing, planning flight paths based on terrain, and identifying targets. . . . Programs such as Dark Sword, Star Shadow, and Sharp Sword are being developed for strike and air-to-air missions."[144] The PLA is experimenting with drone swarming—"to simultaneously watch, jam, and strike high-value targets such as aircraft carriers"—and to overpower American or Taiwanese air defenses in a future war.[145] "The arrival of intelligent drones is expected to accelerate the OODA [observe–orient–decide–act] loop in unimaginable ways and to rewrite the 'rules of the game' of air warfare."[146] In the areas of maintenance and logistics, the PLA is also working to use image recognition to detect the need for maintenance on weapons platforms, and to use big data analysis to increase the efficiency of logistics. In sum, China is investing heavily in AI for both economic and military competition. It has already emerged as a key player in both realms.

Quantum technologies also have significant military applications. Quantum communications can protect military data and communications against cyberattack, aid in search and navigation, and transform encryption and decryption. For example, traditional radar or LIDAR (light detection and ranging) emits radio waves or light pulses, respectively, to locate adversary ships, weapons systems, and other targets and calculate an object's speed and distance. Quantum sensing dramatically improves measurement accuracy. A quantum radar or LIDAR system can perform the same detection with fewer emissions ("allowing for better detection accuracy at the same levels even for stealth or low observable aircraft"), and at lower power levels, which makes these sensors much more difficult for the adversary to locate and jam.[147] The entangled quantum particles "measure everything from photons to disturbances in magnetic or gravitational fields," thus providing dramatically better imaging and sensing that "can be used to help find underground bunkers or submarines hiding in the depths of the ocean."[148] As Kania and Costello summarize, "The realization of quantum radar, imaging, and sensing would enhance domain awareness and targeting, potentially undermining U.S. investments in stealth technologies or even allowing for the tracking of submarines."[149]

China is competing hard against the United States in quantum technologies broadly, and has shot into the lead in quantum communications. In 2017, a Chinese satellite, Mincius, set up a quantum-encrypted communication link between China and Vienna. Many observers characterized this as a "Sputnik moment" in which superior Chinese technology rattled the United States.[150] Since the Mincius launch, China has set up "a 2,000 km quantum-secure communication ground line between Beijing and Shanghai and plans to expand the line across China."[151] China and Russia, as part of their growing military partnership, have established "unhackable" quantum communications over a distance of over 3,500 kilometers.[152] In sum, in addition to

China developing cutting-edge commercial technologies, China's government is exploring the military applications of those technologies. In the critical fields of AI and quantum science, China is recognized as a peer or even as surpassing the United States' technological sophistication.

Counterarguments

Critics might object to this chapter's conclusions for at least two reasons. First, skeptics of China's innovation capabilities have argued that observers too often emphasize a country's inventive capabilities, while overlooking the critically important dimension of diffusion. Second, other critics might argue that China became an innovation leader in highly favorable global and strategic conditions that no longer obtain.

CAN CHINA DIFFUSE INNOVATION?

Some observers continue to see problems with Chinese innovation, which makes them skeptical about China's economic and military competitiveness.[153] Jeffrey Ding argues that scholars overemphasize invention-related activities in their analyses of innovation and argues for the significance of diffusion in the innovation process. Ding contends that China's diffusion capabilities significantly lag those of the United States.[154]

Findings from this chapter challenge this pessimism for a few reasons. First, this chapter conceptualizes innovation broadly—to include diffusion—and measures it through triangulation. The metric of high-tech exports captures the diffusion dimension; I argue that China's success on this metric shows that China is indeed diffusing innovation throughout its economy.

Second, Ding's metrics likely underestimate China's diffusion capabilities for two reasons. Ding differentiates invention versus diffusion, and assigns human capital metrics to measure the former rather than the latter. But human capital is also a key driver—according to economists, *the* key driver—of successful diffusion.[155] And human capital is a metric on which China performs quite well. Bringing human capital back in would create a more optimistic assessment of Chinese diffusion capabilities.

Third, metrics of diffusion expressed at national levels obscure profound regional heterogeneity within China. Ding rightly argues that information, communication, and technology (ICT) are important factors that drive innovation diffusion, and he compares Chinese and US ICT statistics. But national-level statistics obscure significant regional differences. Provincial-level metrics show that China scores high on ICT measures in Beijing, Shanghai, and its dynamic Pearl River Delta region, while scoring lower (reflecting slow diffusion) in its less-developed interior.[156] In other words,

China consists of dynamic, high-diffusion areas—which contribute significantly toward its economic dynamism—as well as low-diffusion areas. National-level statistics obscure this variation to suggest one large mediocre area. For all these reasons, China will likely be more adept at diffusion than Ding and other skeptics argue.

INNOVATION IN THE NEW COLD WAR

Critics might also counter this chapter's optimism about Chinese innovation by arguing that China's innovation success occurred under favorable international conditions that have since soured. The supportive US engagement policy toward China since the 1990s connected China to educational and S&T networks, overseas markets, FDI, and cutting-edge technology that helped China climb up the technology ladder.[157] However, US threat perception of China has risen and Washington has shifted toward a more competitive approach, prompting arguments about a "new cold war."[158] In its first term in office, the Trump administration enacted export controls on Huawei for violating US sanctions against Iran; it imposed tariffs on a number of Chinese products, citing damage to the US economy from Chinese government subsidies and IP theft. US leaders became increasingly concerned about China's ability to access "dual-use" technology that makes up advanced weaponry. Through export controls coordinated with US allies that play key roles in the global semiconductor supply chain, the Biden administration sought to deny China access to cutting-edge semiconductors and related technology. Such an effort was possible because of Chinese dependence on a global supply chain dominated by the United States and its partners.[159]

Worsening relations with the United States and its partners could indeed negatively affect the future of Chinese innovation. If China sends fewer students and professionals overseas and participates less in global S&T networks, this would weaken Chinese S&T capabilities. Furthermore, US export controls threaten to cut off China from key AI inputs and weaken its ties to its closest AI partners. As Brummer and Lind argue, China's key technology partners are some of the countries cooperating with Washington's export controls.[160] If these countries continue to cooperate, this would deny China access to advanced AI inputs and venture capital, as well as to cutting-edge knowledge and research networks. Brummer and Lind show that China currently enjoys strong ties to the world's most innovative AI network, one that includes Japan, the Netherlands, South Korea, the United Kingdom, and the United States.[161] Diminished connections with this network—given the importance of networks in innovation, and the less innovative countries to which China would likely turn—could thus undermine China's future innovation performance.

China certainly faces a more challenging international environment, but this pessimism is probably overstated for a few reasons. A first reason is that China's centrality in the global economy means that countries are loathe to decouple from China because doing so imposes significant costs on their own firms. As such, South Korea and Taiwan sought waivers from US export controls to permit their firms to continue their trade with and operations in China.[162] Furthermore, Chinese firms continue to benefit from integration in global research networks, via state key laboratories as well as research institutes created by private firms around the world. Huawei, for its part, is investing $300 million a year for the next decade into overseas facilities focused on basic science in information and communications technology.[163] As Lind and Mastanduno note in the case of a Huawei chipmaking equipment R&D center, Chinese firms building research facilities at home also recruit former employees of cutting-edge global firms.[164]

The argument that the export controls will keep China off of the technological cutting edge is also problematic. Chinese AI firms have already found ways to work around the export controls. Chinese firms can access embargoed chips through smuggling or theft; they can obtain them through dummy companies; they can rely on cloud computing; they can rent (rather than buy) cutting-edge chips to train their AI models.[165] Furthermore, Chinese firms adapted by using previous-generation chips and lithography equipment. One analyst noted, "You've got to spend $20 million instead of $10 million to train it. . . . Does that suck? Yes it does. But does that mean this is impossible for Alibaba or Baidu? No, that's not a problem."[166] Indeed, notable Chinese advances in both AI models and in semiconductors (e.g., achievements by DeepSeek, Butterfly Effect, and Huawei) raised doubts that export controls are meaningfully holding China back—and suggested they may actually be galvanizing innovation.[167] For all these reasons, export controls may not erode Chinese technological capabilities as much as Washington hopes.

In general, the extent to which other countries are in fact decoupling from China—and thus the extent to which China is becoming increasingly isolated from global educational, business, and technology networks—is overstated. The United States and China remain major trading partners, with total bilateral trade totaling $582.5 billion in 2024. Export controls notwithstanding, the US Department of Commerce approves most licenses sought by US exporters—as critics of the Biden administration lamented.[168] Chinese trade also remains robust with other partners; beyond the narrow area of <7-nanometer chips covered by the export controls, China continues to be the number-one trading partner for Japan, South Korea, Taiwan, and many other countries. To be sure, China would likely enjoy a more favorable position had its policies not antagonized the United States and some of its neighbors. However, with China possessing a vast, diverse internal market, led by a government that is spending hundreds of billions of dollars to promote indigenous innovation, and still enjoying extensive global ties and influence, one can make a good case that Chinese innovation will remain strong.

A Matter of Life or Death for the Nation

A great power's need to perform at the technological cutting edge—and the dire implications of failure—are starkly clear in the example of the Soviet Union. A technological leader in the second Industrial Revolution, the USSR fell behind, unable to keep up economically or militarily, after the advent of the information age. Eventually the Soviet government fell in a coup d'etat; many republics stampeded out the exits, leading Russia to lose a quarter of its former territory and half its population. Its geopolitical adversary then pushed into Russia's former territory and sphere of influence. The USSR's decline highlights how technological competitiveness is, as Chinese leaders observed, "a matter of life or death for the nation."[169]

Can China compete at the technological frontier? This chapter argues yes: that in addition to having the massive scale of a great power, China has become an S&T leader. China is not the world's top performer, and indeed may never overtake the United States. China is still developing its "innovation system"—for example, norms of academic integrity, business-government relations, property rights, its regulatory environment.[170] And as much as the CCP has invested in education, China still faces the daunting challenge of educating its massive rural population.[171] But the key question is whether China is competing at the global technological cutting edge—and this chapter shows that it is. Furthermore, trends suggest that China's S&T excellence is likely to endure: that China rivals the United States as a world leader in the technologies of the 4IR.

"I MADE THE TEA"

Why did observers in the United States and the West so underestimate China's innovation potential? It's worth noting that this not a new phenomenon in great-power rise.[172] When the British began colonizing North America, wrote Doron S. Ben-Atar, "no imperial statesman envisioned that these struggling outposts could become actual economic rivals."[173] Later, many in the United States dismissed Japan's economic potential. In 1954, US Secretary of State John Foster Dulles "told Premier Yoshida frankly that Japan should not expect to find a big American market because the Japanese don't make the things we want" and thus "Japan must find markets elsewhere for the goods they export."[174] Great powers, it seems, tend to underestimate newcomers. Beyond this tendency, Western observers likely underestimated China because they conceived of innovation quite narrowly (as invention), and because of US distrust of a heavy state role in innovation.

A manufacturing powerhouse central to global supply chains, China excelled at process innovation. China developed, wrote Breznitz and Murphree, "a formidable competitive capacity to innovate in different segments of the research, development, and production chain that are as critical for economic growth as many novel-product innovations, and perhaps even more so."[175] Dan Wang argued that although China has made "relatively modest contributions to pathbreaking research and scientific innovation," the country has excelled in the "less flashy task of improving manufacturing capabilities."[176] Chinese firms have innovated in many sectors, argues Wang—in particular, dominating world markets in solar panels and batteries—by "driving costs down through more efficient production." McKinsey Global Institute (MGI) researchers similarly distinguished between different types of innovation, noting that China excels in process innovations because of "the vast scale of China's manufacturing ecosystem." MGI noted that "China has more than five times the supplier base of Japan, 150 million factory workers, and modern transportation." This ecosystem, argues MGI, "enables process innovations that can cut cost, improve quality, and improve flexibility."[177]

MGI also praised China for its achievements in customer-focused innovation, in which people adapt products, services, and business models to different markets and different functional areas. This type of innovation most dramatically illustrates how the drivers of innovation vary across types; in contrast to science-based innovation, customer-focused innovation requires small capital infusions and is performed both by entrepreneurs and by large firms. Customer-focused innovation thrives in large markets with significant regional variation and global market access—conditions that China provides. MGI argues that with a massive market of 1.3 billion people, China "gives innovators a huge supply of problems to solve and needs to fill." Companies also benefit from China's engaged consumers, who are "happy to accept products that are not completely refined and eager to share feedback to make them better."[178] Discussing Chinese innovation in mobile payments and the broader "fintech" sector, Zak Dychtwald sees China's "hundreds of millions of hyper-adoptive and hyper-adaptive consumers" as part of its "new innovation advantage."[179]

A second reason why US observers in particular underestimated Chinese innovation relates to its state-directed model. At the core of the US national identity, and the rationale behind US political institutions, is a belief about the dangers of a powerful central government. This identity contributes to a pervasive narrative in the United States about the evils of government intervention in the economy. Linda Weiss writes that the United States' innovation system is commonly perceived as "state-less" and "free-wheeling"; that US technological leadership "is attributed to a culture of risk-taking and entrepreneurship in which creative individuals, working on their own initiative, push out new ideas and new widgets based on their own ingenuity and derring-do."[180] In this view, innovators are tinkerers in garages who need to be connected with venture capital; the less the government gets involved, the better.[181] Scholars (correctly) make arguments about the dangers of state intervention and cite numerous examples in which an overbearing, misguided, or captured government quashed rather than sparked innovation.[182]

The bias against the role of the state in economic policy (particularly pronounced in the United States) is problematic for a few reasons. First, it rests on an inaccurate narrative about the roots of US technology dominance. US technology leadership was not only rooted in the creativity of garage tinkerers, but in its "stealth industrial policy,"[183] what Weiss calls the "national security state—a particular cluster of federal agencies that collaborate closely with private actors in pursuit of security-related objectives." The national security state, Weiss argues, "is geared to the permanent mobilization of the nation's science and technology resources for military primacy."[184] Even in the free-wheeling United States, then, the government played a highly influential role in fostering technological innovation.

Furthermore, the evergreen debate about the state's role in the economy is by no means resolved. Cases of predatory, poorly conceived innovation policies can be rebutted with examples of stunning successes. In South Korea and Taiwan, for example, governments provided significant support to create a semiconductor industry in the 1980s and 1990s, leading those two countries to become global leaders.[185] And in China itself, strong direction and heavy investment from the CCP have paid off in several sectors, notably solar panels and EV batteries.[186] Ultimately, as Breznitz argues, countries fostered innovation through a variety of strategies, many of which featured a prominent role for the state.[187]

In sum, pessimism about Chinese S&T capabilities is often rooted in prejudice against state-centric models, and in a definition of innovation that emphasizes science-driven invention. But China has excelled in process-driven innovation, and importantly, Chinese investments in basic science—both by the CCP and by Chinese firms—are starting to pay off too.[188] China may never catch up to the United States, which remains on most metrics the world's most innovative country due to a long list of qualities (e.g., ample capital, centrality to both the Asian as well as European innovation networks, a world-renowned university system). Critically, however, great powers need not catch up to the leading state in order to represent a dangerous geopolitical and military threat; rather, great powers need to operate at the global S&T cutting edge. Indeed, as this chapter has shown, China has already caught up to and even surpassed several countries that are considered top players in terms of their innovative capabilities.

Finally, many observers doubted China's innovation potential because of its authoritarian institutions. Although scholars recognize that authoritarian countries sometimes mobilize inputs to generate catch-up growth, many observers—drawing on a prominent literature about institutions in economic development—argue that in the modern era, authoritarian regimes cannot provide the conditions necessary to foster an innovative economy.[189] China's performance at the global S&T frontier raises a puzzle for this widespread view. The following chapters explain this puzzle with an argument about "smart authoritarianism."

CHAPTER 3

Smart Authoritarianism

J. B. Jeyaretnam sold T-shirts. Over his long career, the Singaporean opposition politician faced several government lawsuits (for libel, defamation, embezzlement, and so on). Bankrupted by court fines, he was ineligible to run for Parliament until he paid them off. In March 2001 "JBJ," as he was called, held a rally in Yio Chu Kang Stadium, hoping to raise $400,000 that would enable him to stand for election. "Power doesn't belong to the government," Jeyaretnam told the crowd of 5,000. "It belongs to the people."[1] Supporters of his Worker's Party sold stickers, copies of his book (*Make It Right for Singapore*), and T-shirts that displayed the message "We the People." Scholar Michael Barr, who was in the crowd that evening, had doubts about the plan. "I recognised several people in the crowd as beggars," he commented. "This is hardly the crowd to produce half-a-million dollars in profit from buying bumper stickers and T-shirts."[2] Indeed, the Worker's Party made $20,000. JBJ lost his seat.

While it was perhaps not surprising that the rally failed its fundraising target, what is surprising about JBJ's career is that he had one at all. In other authoritarian countries, opposition leaders end up in the back of a black van in the middle of the night, in a detention center interrogated and beaten by security forces, or lying dead in a ditch. JBJ's story is intertwined with the story of authoritarian adaptation; smart authoritarian leaders are the smooth, Savile Row–clad progeny of yesterday's brass-knuckle dictators. Their new styles of authoritarian rule enabled China, as Chapter 2 showed, to successfully cultivate science and technology (S&T) innovation, leading it to become one of the world's most innovative countries and a superpower competitor to the United States. How?

This book argues that autocratic regimes can stay in power and foster innovation by adopting a model of "smart authoritarianism." This chapter lays out the argument. It explains how authoritarian regimes preside over not only catch-up growth (as seen in the cases of South Korea's and Taiwan's "developmental dictators") but also innovation-based growth. The argument agrees with assumptions in institutions theory about tensions between economic growth and political control, and it expects that these tensions are strongest at the stage of innovation-based growth. But I argue that institutions theory errs in seeing these tensions as irreconcilable, and that the theory neglects significant variation among autocracies across both time and space.

Namely, autocracies around the world adopt very different policies—to put it in terms of institutions theory, the "extractiveness" of their institutions varies. Many observers (and indeed a large international relations [IR] literature) put countries into one of two buckets—democratic or authoritarian—arguing that countries in the latter bucket (with extractive political institutions) inevitably adopt extractive economic institutions. But authoritarian countries govern with very different types of institutions and adopt diverse economic policies—with profound implications for economic growth and innovation.[3]

Furthermore, authoritarian regimes have changed significantly over time. They have adapted along with changes in the global economy since the late twentieth century. As noted in earlier in the book, Aristotle once advised rulers to be "not vicious but half-vicious": to make themselves more accepted by the people, and their rule more durable as a result.[4] Smart authoritarians listened; to foster economic growth and innovation they moderated the viciousness of their political institutions and adopted inclusive economic policies. The essence of smart authoritarianism is a regime's effort to maintain a delicate balance between maximal political control and the minimal rights and freedoms necessary to foster economic growth and innovation.

In this chapter, I first review the institutions argument and note empirical and logical problems with the claim. Next I lay out the logic of smart authoritarianism. I show how smart authoritarian regimes provide the conditions (discussed in Chapter 1) that support innovation—for example, property rights, capital, and a civil society. I conclude by addressing counterarguments and discussing the conditions under which scholars expect authoritarian governments to adopt "smart" policies.

A quick note on adjectives. In this chapter I adopt the terminology "smart" authoritarianism to refer to autocratic governments that seek to foster economic growth and innovation by calibrating policies of authoritarian control with inclusive economic policies. "Smart" suggests approbation. And indeed, many people celebrate developmental dictators—for example, Singapore's Lee Kuan Yew, South Korea's Park Chung-hee, Mexico's Porfirio Diaz, Turkey's Mustafa Kemal Atatürk—as nation-building heroes. But it's important to remember the human cost of the repression that accompanied their developmental policies. People died. Political prisoners suffered. People who loved their countries and wanted to make them better—politicians, journalists, teachers, writers, activists—were driven from their fields, or even their countries. Any praise for the achievements of smart authoritarianism must remember and grapple with these human costs. Legal scholar Jothie Rajah asks unflinchingly: "Does the delivery of employment, infrastructure, and social order," she asks, "make for some sort of realpolitik balance sheet in which the political violence visited upon a few is set off against general contentment?"[5] "Smart" authoritarian policies may have indeed been less bloody relative to other authoritarian alternatives, but they were still part of a system of repression. Even half-vicious is still vicious.

Furthermore, my students sometimes jokingly refer to the opposite of smart authoritarianism as "dumb" authoritarianism. But leaders who maneuver past millions of people to rule their countries, and sometimes hold power for decades, are far from dumb. Ruthless perhaps, but not dumb. Dictators who overvalue exchange rates and create import quotas understand that such policies harm the economy, but those leaders also understand that such policies yield rents that make sense for a regime under some conditions. Moving toward smart authoritarianism requires change and risk, which not every dictator may deem necessary or wise. Its opposite (which I prefer to call "blunt" authoritarianism) may be perfectly rational, however loathsome, for leaders who continue to rule—and may even successfully stay in power—through policies of isolation, corruption, and conspicuous brutality. The end of this chapter explores conditions under which authoritarian leaders are more likely to adopt the smart model.

Innovation and the King's Dilemma

As described in Chapter 1, great powers must operate at the global S&T frontier. Innovation is essential both to sustain economic growth and to develop advanced weaponry, without which great powers will struggle to compete. According to institutions theorists, authoritarian regimes rule through extractive institutions, so are unable to provide the conditions that foster innovation (e.g., property rights, high-quality human capital, networks and connectivity). Late twentieth-century trends (such as the advent of the information age and the rise of a liberal development regime) created a "web of constraints" that advantaged liberal states in economic development.[6]

In this view, authoritarian rulers face a dilemma—a version of Samuel Huntington's "king's dilemma."[7] A regime could maintain the extractive institutions that kept it in power—leading to an underskilled workforce and a dearth of capital due to unpredictable economic conditions and international opprobrium. This would create economic and technological mediocrity, making the country unable to compete geopolitically against more liberal countries. Economist Arthur Kroeber argued that "it is impossible to imagine creativity blossoming as long as the state places draconian restrictions on the right of people to express their views, share information, organize independently to solve social problems."[8] Alternatively, an autocracy could reduce repression, invest in human capital, refrain from predatory economic policies, open up its economy, permit its people greater freedoms, and allow the free flow of information. This would promote economic growth and innovation—but would unleash forces for liberalization that would ultimately topple the regime. As G. John Ikenberry contends, "There is little evidence that authoritarian states can become truly advanced societies without moving in a liberal democratic direction."[9]

In short, "Dictators face a dilemma," argues Matthew Kroenig. "They can put in place policies that encourage economic growth only by threatening their own powers. Or they can opt for suboptimal economic performance and the protection of their privileged productive sectors, but these models of growth have limits."[10] According to the institutions view, authoritarian leaders have to choose between staying in power or the technological capabilities that drive great-power competitiveness. In the words of a Chinese saying, "Loosening causes chaos; tightening up causes death."[11]

Smart Authoritarianism

The smart authoritarian argument shares several assumptions with the argument described above. Countries must cultivate strong national S&T capabilities in order to compete in international politics—particularly against great powers. Authoritarian regimes (currently) face a tension between providing the conditions to foster technology innovation on the one hand and staying in power on the other hand.[12] Furthermore, trends in technology and international politics since the late twentieth century have exacerbated this tension, to the disadvantage of autocratic states. However, according to the smart authoritarian model, the institutions argument errs in seeing an irreconcilable dilemma between control and innovation: in viewing authoritarian regimes as homogenous and static.

THE DIVERSE AUTHORITARIANS

The institutions argument overlooks significant heterogeneity among authoritarian regimes and conflates institutions with the policies that drive the economy. The argument separates countries into two buckets: democracies and autocracies. Because of the extractive institutions that keep a regime in power, autocracies are seen as unable to provide the conditions that foster innovation.

Chapter 2 has already discussed one problem with this claim, which is that it lumps together innovation when in fact there are many different types of innovation with different drivers.[13] In other words, perhaps authoritarianism is ill-suited to cultivating some types of innovation (e.g., consumer-focused or engineering-based innovation?) but is actually capable of—perhaps even advantaged at—fostering other types.

The institutions argument also lumps together authoritarian regimes, neglecting to see variation in the extractiveness of a regime's political institutions and assuming that countries with extractive political institutions inevitably adopt extractive economic institutions. But in fact authoritarian regimes are heterogenous both in their political institutions and in the economic policies they adopt. Although this is well-trodden territory for scholars of authoritarian politics and political economy, the IR literature has not yet digested this fact.[14] Indeed, a large IR literature on "democratic difference" focuses on the distinction between democracies and autocracies while neglecting significant variation in the latter.[15]

Many scholars emphasize that a regime's policies, not its institutions, drive economic outcomes.[16] The "adoption of policies consistent with economic freedom—greater reliance on markets, freedom of exchange, openness of the economy, and monetary stability," argue James Gwartney, Robert Lawson, and Randall Holcombe, "is more important as a source of economic growth than the nature of the political regime."[17] Analysts often have a tendency to equate political institutions with economic ones (e.g., associating "property rights" or "central bank independence" with democracy), but in fact these are inclusive economic practices that either democracies or autocracies may adopt.[18] Glaeser et al. argue that any leader may choose to respect property rights, invest in human capital, and so on: "Promarket dictators can secure property rights as a matter of policy choice, not of political constraints." These authors point out that in the 1980s, two of the world's top ten countries with the lowest expropriation risk were Singapore and the USSR. "Whatever expropriation risk measures," they note, "it is obviously not permanent rules, procedures, or norms supplying checks and balances on the sovereign."[19] They conclude that "the economic success of East Asia in the postwar era, and of China most recently, has been a consequence of good-for-growth dictators, not of institutions constraining them."[20] Extractive political institutions, in other words, do not necessarily produce extractive economic institutions.

Indeed, around the world, authoritarian economic policies vary dramatically. Argues Joseph Wright, "We are beginning to understand that variation among different types of authoritarian polities can perhaps be as important as the distinction between democracies and dictatorships."[21] Erich Weede notes that authoritarians diverge significantly in their economic policies "either in the direction of efficiency and growth promotion, like Lee Kuan Yew in Singapore, or in the direction of kleptocracy, like Mobutu in Zaire."[22] Following the pathbreaking work of Barbara Geddes, scholars distinguish between different kinds of authoritarian regimes (e.g., "personalist," "single-party dictatorship," and so on) and study how different types perform on various dimensions (including economic growth).[23] Authoritarian heterogeneity helps explain why scholars continue to debate empirical support for a democratic advantage in growth. Argues Stephan Haggard, "One reason why the cross-national evidence on the relationship between regime type and performance is so weak and contradictory is that the 'authoritarian' category is so diverse."[24]

AUTHORITARIAN ADAPTATION

So, autocracies are not homogenous—and neither are they static. The institutions argument expected that as late twentieth-century trends disadvantaged autocracies in economic growth, regimes either would relax constraints and fall from power—or would maintain a tight grip on power, which would stifle economic growth. In the words of the Chinese maxim, a regime faced chaos or death. But smart authoritarians chose Door #3; they adapted.

William J. Dobson argues that "dictators are far more sophisticated, savvy, and nimble than they once were": "refashioning dictatorship for the modern age."[25] A large literature examines the emergence of new forms of authoritarianism called, variously, "hybrid regimes," "delegative democracy," "illiberal democracy," "soft authoritarianism," or (my personal favorite) "vegetarian" authoritarianism.[26] Authoritarianism, like COVID-19, evolved from a delta variant to what may in fact be a more effective omicron: what I call smart authoritarianism.

Scholars have demonstrated a general trend in the late twentieth century, across a wide range of issues, in which autocracies tactically changed their behavior in response to constraints imposed by wealthy liberal countries and the international order they created. In response to aid conditionality and the growing significance of global rankings, autocracies adapted their policies in order to attract foreign aid and FDI.[27] They tactically participated in international justice tribunals, sought ascension to the European Union and other international institutions, accepted election monitoring, and created gender quotas for elected office: painting a patina of liberalism to please liberal observers.[28] As Chapter 5 demonstrates, authoritarian regimes engage in "sportswashing" (think the Sochi 2014 Winter Olympics and Saudi Arabia's LIV Golf League) and other kinds of tactics to manage their image in the eyes of the global community. Below I explore the mechanisms of authoritarian adaptation—and how autocratic regimes are able to provide the conditions that foster innovation.

THE SPACE BETWEEN "CHAOS AND DEATH"

Smart authoritarian regimes understand the drivers of innovation in the modern global economy: highly skilled workers, information flows, openness to the world, and so on. But those leaders also understand that the forces that make a society more innovative threaten the regime's political control. Smart authoritarian leaders thus want to position the society in the space between maximal regime control (to stay in power) and minimal openness (to foster innovation).

As such, smart authoritarianism is not a growth-maximizing strategy. The strategy seeks to create a competitive economy while maintaining regime control. Smart authoritarian leaders understand that by following this strategy they are leaving some economic growth on the table—that is, that with more openness and freedoms, the country could have been even more innovative and could have grown more rapidly. But that "delta" growth could come at the cost of the regime losing power. Thus the regime's goal is not growth maximization, but rather growth *and* political control.

Many observers, for example, reacted to Xi Jinping's tightening of political controls over the economy with puzzlement—how could such an ambitious, shrewd leader be so misguided to enact policies that would slow his country's economic growth? Speculation ensued: Perhaps Xi had become drunk on power; perhaps he had walled himself off from good information by surrounding himself with yes-men. Such arguments are an effort to explain behavior that people find irrational: why a regime would ever enact policies that would reduce economic growth.

Perhaps some authoritarian leaders are indeed drunk on power, surrounded by yes-men, and so forth. And perhaps someday scholars will conclude that Xi was one of them. But another explanation for why highly shrewd leaders tighten controls over the economy is that they are following the smart authoritarian model—seeking to have their economic growth and stay in power, too.

Smart authoritarian leaders, like sailors, have to constantly tack between more control and more growth. At times they'll loosen the sails, seeking greater economic growth and innovation. Just as sailors are constantly glancing at their wind indicators, smart authoritarians keep a close eye on economic, societal, and technological trends. At times when the regime perceives growing threats, leaders will tighten the line, accepting less growth and innovation. As Curtis Ryan observed of Jordan's authoritarian regime, "Jordan signals constant movement and micro-levels of change, in order to essentially stay the same."[29]

As authoritarian leaders seek to find that sweet spot between authoritarian control and economic growth, they face constantly changing conditions. First, the sweet spot will change as the economy develops; regimes at the catch-up stage face different challenges for economic growth and for control than regimes trying to foster innovation-based growth. Second, changes in the nature of S&T transform the nature and drivers of innovation. These changed dramatically from the second Industrial Revolution to the third— and are changing again in the fourth.[30] Scholars see artificial intelligence

(AI) as causing a "Kuhnian" or paradigm shift in science akin to previous revolutions such as the theory of evolution and quantum mechanics.[31] AI is changing the way science is performed: "increasingly integrated into scientific discovery to augment and accelerate research, helping scientists to generate hypotheses, design experiments, collect and interpret large datasets, and gain insights that might not have been possible using traditional scientific methods alone."[32] Authoritarian rulers seeking to balance between control and innovation thus face ongoing change in the very nature of science and innovation.

Furthermore, the tools of control available to authoritarian leaders are constantly evolving too. Whereas authoritarian leaders previously suppressed dissent with truncheons and guns, today's dictators draw upon far wider—and subtler—tools of repression, which affect growth and innovation differently. In sum, as their economy advances, as technology and innovation change, and as a regime's tools of control change, the sweet spot between political control and innovation constantly shifts; authoritarian regimes must constantly learn, update, and adjust. They are fortunate in that they are not alone in this challenge. Autocrats support each other and share information about best practices with other regimes around the world.[33]

In sum, institutions arguments see dictators as facing an inexorable dilemma between innovation and control, but smart authoritarians wriggled out of it. Facing disadvantageous changes in the global economy in the late twentieth century, authoritarians did not go gentle into that good night. They calibrated their policies between repressing and controlling enough that the regime stays in power, while providing enough openness and freedoms to cultivate growth and innovation.

Mechanisms of Smart Authoritarianism

Smart authoritarian leaders moderate the extractiveness of their political institutions and, through a range of economic policies, provide conditions that foster innovation.

MORE INCLUSIVE POLITICAL INSTITUTIONS

Authoritarian regimes have adapted to the less favorable conditions of the late twentieth century by moderating the extractiveness of their political institutions. Today's authoritarian regimes, argues Erica Frantz, "differ from their Cold War-era predecessors primarily in terms of the extent to which they seek to mimic democratic rule."[34] More autocrats are ruling through political parties and legislatures, and more are holding elections; some regimes even allow opposition parties (recalling the experience of

Singapore's J. B. Jeyaretnam at the start of this chapter).[35] Such institutions help a regime "disguise their true authoritarian nature behind the façade of democracy," helping it "stay in the good graces of the international community."[36] Holding elections—even those in which the incumbent enjoys numerous advantages—often placates liberal observers' requirements for favorable trade terms, aid, and so on.

Furthermore, more inclusive institutions also serve the regime's interest. Scholars note that institutions such as parliaments help co-opt potential challengers; they "serve as a forum in which the regime and opposition can announce their policy preferences and forge agreements";[37] they reassure members of the selectorate that a dictator will provide promised rents.[38] Dictators can still control access to institutions, and they can control information about negotiations within them.[39] Autocrats are wise to adopt "pseudo-democratic institutions," notes Frantz; the regimes that do "last quite a bit longer in power than those without them."[40]

LOW-INTENSITY REPRESSION

Dictators of the past would often arrest thousands of perceived enemies of the regime, then brutally execute them or send them to gulags or prison.[41] Previous regimes relied on heads-on-pikes showmanship to deter future challengers.[42] "We want to be shocking," declared Mao Zedong, whose campaigns tormented and killed tens of millions of Chinese.[43] In Rafael Trujillo's Dominican Republic, terror was "theatrical": "taken to new extremes."[44]

Today's regimes are different. Michael Ignatieff, Canadian politician and former president of the Central European University (ousted from Budapest by Viktor Orbán's government), calls today's dictatorships "a new thing under the sun." "The closed regimes of the past," he comments, "were behind barbed-wire fences and police watchtowers, and the repression was overt and clear and unmistakable."[45] But today, "a less carnivorous form of authoritarian government has emerged," argue Sergei Guriev and Daniel Triesman; autocrats are getting "more surgical"; they "use violence sparingly. They prefer the ankle bracelet to the Gulag."[46]

Increasingly mindful of global optics, smart authoritarians shifted from high- to low-intensity repression.[47] Low-intensity repression attracts less attention "not only because it is understated, but also because it is often a series of small, individual incidents as opposed to a single, large-scale event."[48] Political opponents are arrested or fined for nonpolitical crimes. Dictators increasingly silence activists and opposition politicians—recall Singapore's J. B. Jeyaretnam—through defamation lawsuits, character assassination, or the denial of rights or benefits. Political opponents are accused of taking bribes, being on the payroll of foreign enemies, or engaging in immoral or sexually deviant behavior.[49] In Malaysia, for example, opposition politician Anwar Ibrahim was "imprisoned on corruption and sodomy charges—a charge specifically designed to delegitimize him in the eyes of a still largely homophobic Malaysian public."[50] Steven Levitsky and Lucan Way describe "more subtle forms of persecution, such as the use of tax authorities, compliant judiciaries, and other state agencies to 'legally' harass, persecute, or extort cooperative behavior from critics."[51] Dobson argues that "law, regulation, and procedure can be a dictator's most effective tools for strangling an opponent." He explains:

> If you seek to disband an NGO, you don't arrest its membership, you send health inspectors to temporarily close its headquarters, pending a review of a series of alleged health code violations. If you are troubled by what a radio station is broadcasting, you don't . . . force it off the airwaves. Rather, you send tax inspectors to audit the station's books and find financial irregularities that require the station's temporary closure. . . . The mere threat of legal sanctions or administrative review may encourage the radio station's management to engage in the very self-censorship that accomplishes the regime's ends.[52]

In sum, smart autocrats silence critics not with bullets but with lawsuits and red tape.

Smart authoritarians avoid the need for repression in the first place because they train the public in the types of expression the regime will tolerate. In Vietnam, for example, authorities "are more tolerant of criticism about particular government policies or programs or of particular nonsenior officials than they are of criticism about top national leaders, the form of government, or the entire political system";[53] protests by individual critics are more tolerated than those by "large groups that publicly rebuke a policy or program." Furthermore, "authorities are more tolerant of protests by peasants and workers than they are of demonstrations by middle class, rather well- educated urbanites."[54] Protestors learn, in other words, what kind of political speech the regime will accept (and by whom), and adjust their behavior accordingly, obviating the need for repression.

Over time, communist countries embraced these subtler strategies of repression. In the Soviet Union under Andropov, describe Guriev and Treisman, "Telephones were disconnected, driver's licenses suspended, typewriters confiscated . . . children denied college admissions"; "Andropov's men preferred frequent searches and interrogations, short-term preventive detentions, and house arrest."[55] The government responded to small infractions not with arrests but with a summons and a frightening talking-to by the KGB, after which people were sent home. Guriev and Treisman argue that the East German government under Erich Honecker surpassed the Soviets in subtlety, with methods that sought "to disrupt the target's personal life and career, isolate him from friends and family, and make him question his own sanity."[56] Today Vietnam follows similar methods. Although the government occasionally uses more forceful measures, primarily its agents "harass regime dissidents and their families." For example, "authorities tap and cut phone lines to dissidents' residences, block or disrupt their mobile phone numbers, hack into their email correspondence, track their internet usage, and confiscate files, books, letters, and computers from their homes."[57] Smart authoritarians thus rely on less bloody and more targeted repression, which relative to high-intensity repression is more conducive to economic growth.

PROPERTY RIGHTS

As described in Chapter 1, encouraging innovation requires solving a key market failure by providing property rights; entrepreneurs and inventors need to know that they will profit from their risk, cost, and labors. Although authoritarianism has long been associated with weak property rights, smart authoritarians abstain from expropriation. Adam Przeworski argues that economists often use "risk of expropriation" as a measurement of institutional constraint, but a decision to expropriate is in fact a choice made by an unconstrained leader.[58] While personalist rulers often expropriate, other authoritarian leaders are more likely to respect property rights in order to encourage investment.[59]

A well-functioning property rights regime requires a judiciary, which smart authoritarians also create. Institutional checks and balances create "horizontal accountability," and the democratic advantage school does not expect authoritarian leaders to accept such constraints on their power.[60] But other scholars remind us that "authoritarianism and the 'rule of law' are not mutually incompatible"; and indeed that "the rule of law ideal initially developed in non-liberal societies."[61] Scholars point out that judiciaries in fact benefit authoritarian regimes in a number of ways. Courts encourage investment by reassuring investors that they can adjudicate property rights, thus promoting economic growth and innovation. Additionally, as Tamir Moustafa has written, the courts

> (1) establish social control and sideline political opponents, (2) bolster a regime's claim to "legal" legitimacy, (3) strengthen administrative compliance within the state's own bureaucratic machinery and solve coordination problems among competing factions within the regime, (4) facilitate trade and investment, and (5) implement controversial policies so as to allow political distance from core elements of the regime.[62]

Judiciaries promote political stability, too. Jacqueline Sievert finds that "states that adopt an independent court reduce their risk of civil war between 54 and 75 percent."[63] Dictators are getting the message; Sievert notes that whereas in 1961 only 9 percent of autocracies had partially or fully independent courts, by 1987 more than a third had pursued judicial reform.

How to enjoy the benefits of a judiciary without it constraining regime power? Yuhua Wang argues that dictators create a partial rule of law and sequence their legal reforms: In many cases, leaders "'tie their hands' in commercial cases yet extend their discretionary power in the political realm."[64] Gordon Silverstein argues that "while economic reform and prosperity demand the rule of law, the rule of law does not necessarily mean that judicialization—and the expansion of individual rights—necessarily will follow. It is possible to de-link economic and political/social reform."[65] In other words, an absence of democracy does not equal an absence of the rule of law. Smart authoritarian governments protect property rights and create judiciaries in order to reassure investors and encourage growth and innovation.

CAPITAL

Smart authoritarians create stable macroeconomic conditions to encourage the capital formation and investment required for innovation. In part they do this by building the types of institutions (e.g., parliaments, judiciaries, central banks) described earlier. Such institutions reassure investors by "increase[ing] the transparency of the decision-making process" and encouraging "stable investment environments."[66]

Smart authoritarians also reassure investors by avoiding some of the most economically distorting practices common in other authoritarian regimes. In order to fund the patronage that they need to stay in power, many authoritarian leaders spend lavishly,[67] interfere with monetary policy,[68] default on debt,[69] and create multiple exchange rates (in order to bestow favorable ones on supporters).[70] Smart authoritarians, by contrast, avoid these economically destabilizing practices. For example, authoritarian South Korea had "a relatively stable macroeconomic environment characterised by limited inflation relative to many developing countries": "Rarely did the real effective exchange rate appreciate and such episodes were quickly corrected."[71] Michael Rock notes that in Indonesia, Malaysia, and Thailand, core economic agencies "maintain[ed] macroeconomic stability and a competitive exchange rate."[72]

In order to provide the macroeconomic stability necessary to attract FDI and encourage capital formation, smart authoritarian regimes create a professionalized bureaucracy. The institutions view expects the civil service in authoritarian countries to be rife with nepotism and patronage—and in many countries it is.[73] Smart authoritarians, however, create a professionalized civil service. Through "bifurcation" or "pockets of excellence," autocrats shelter core macroeconomic agencies (e.g., central banks, finance ministries, and national planning agencies) from the patronage that leaders need to reward their political allies. In those key offices, leaders install a professional, competent cadre of macroeconomic managers.[74] (Patronage positions can still be doled out in other agencies—for example, the Ministries of the Interior, Transportation, or Construction.)

In this manner, East Asia's developmental dictators created professionalized civil services in their pursuit of catch-up growth.[75] In the United Arab Emirates (UAE), Mohammed bin Zayed (MbZ)'s economic reforms "began with Abu Dhabi's Civil Service, which was afflicted with many of the same ills as those of other Arab countries: bloat and inefficiency, with connections and family reputation playing a bigger role in hiring than merit." The reformist sheikh "deployed a group of young, talented people and authorized them to smash up the bureaucracy"; they slashed the size of the government from 64,000 employees to 7,000.[76]

Strategies of bifurcation are actually common across all regime types. In the United States, presidents traditionally make some appointments based on political or fundraising considerations but take care to appoint highly qualified experts to vital posts.[77] Richard Katz describes a "dual economy" in Japan's post–World War II economic miracle. Japan, Katz argues, exhibited "a dysfunctional hybrid of super-strong and super-weak sectors" in which the tradeable sector was efficient and meritocratic, and the nontradable sector and ministries were rife with corruption.[78] Ultimately, as Elliot Green notes, "the level of democracy as measured by Polity IV and others may not be the most important factor in explaining patronage distribution, thereby adding to a growing literature that suggests that democracies and nondemocracies in the developing world have fewer policy differences than previously thought."[79]

HIGH-SKILLED HUMAN CAPITAL

Many policy choices affect the quality of a country's workforce. A government can choose to spend more or less on education; institutions theory contends that dictators will focus spending on their selectorate while neglecting public goods. Second, educational quality depends on the competence of the bureaucracy that manages it; as discussed, many scholars expect that patronage creates incompetent and corrupt bureaucracies that waste education (and infrastructure and public health) budgets. Third, governments can decide to control educational content—requiring education that features rote memorization, specialization, and "no-go" zones; alternatively, schools may be empowered to adopt curricula that encourage creativity and critical thinking. Finally, a government affects the caliber of human capital through the country's level of openness—information flows and opportunities to study and work abroad with researchers at the global technological frontier.

Smart authoritarians create high-quality human capital through their policy choices across all of these categories. In the East Asian cases, developmental dictators spent heavily on public goods, instituting universal education and improving access to tertiary education.[80] In the UAE, Zayed bin Sultan al-Nahyan—father of MbZ—instituted universal education when illiteracy among women was nearly 100 percent.[81] Furthermore, as noted, developmental dictators create professionalized bureaucracies to manage their education systems. As for style of education, democracies and autocracies alike often elect to foster an education system that relies heavily on rote memorization and a sequence of examinations (e.g., the "exam hell" system in Japan—one of the world's most innovative countries). Finally, as discussed below, smart authoritarians permit their students to study abroad, as those students will bring home knowledge and ideas that will lift the country's technological base.

FREEDOMS AND CIVIL SOCIETY

Smart authoritarians allow a (government-controlled) civil society that encourages information flows, improves governance, and enables people to explore their interests. At the same time, smart authoritarians impose significant government oversight over civil society organizations and their activities, requiring official permission to operate and monitoring for anti-regime speech and activity.[82]

Authoritarian rulers allow private media, a private business sector, universities, and myriad civil society organizations (CSOs). Governments try to channel citizen engagement into "activities that are deemed civic, 'cultured,' and civilized so that they will keep a safe distance from activities such as political lobbying, protests, campaigning, and even politically induced violence."[83] Governments expel foreign NGOs or control them through bureaucratic harassment and "foreign agent" laws.[84] Writes Frantz, smart authoritarians are "letting nongovernmental organizations operate but secretly requiring them to promote the government line."[85] Governments also operate government-organized nongovernmental organizations (GONGOs). These "typically profess to be independent entities and may hide behind innocuous-sounding names that suggest that their chief mission would be human rights, legal reform, or the protection of minorities. In truth, their goal is to legitimize government policy, soak up foreign funding from genuine NGOs, and confuse the public about who is in the right, the government or its critics."[86]

Smart autocrats also allow the rise of private media, which benefits a regime in important ways. As Georgy Egorov, Sergei Guriev, and Konstantin Sonin describe, private media play an invaluable role in authoritarian regime resilience.[87] Dictators need to know if their bureaucracies and local officials are competent, because their incompetence may fuel public dissatisfaction and antigovernment activity. Yet complete media freedom might publicize government failures and allow the expression of antiregime activity. In order to gain the benefits of monitoring officials while evading the potential costs of public mobilization against the regime, autocrats allow a private media but impose strict limitations upon it.[88]

Smart authoritarians use a variety of strategies to control information. The traditional tool is censorship. Censorship guidelines instruct producers, editors, and reporters about the boundaries of permissible coverage—and over time these professionals learn to self-censor in order to avoid problems. In Venezuela, the Hugo Chávez regime closed 34 radio stations while announcing its investigation of 240 other media outlets—without identifying them. "With the threat of closure already made real," notes Dobson, "the government knew the stations would do its censorship for it."[89]

Increasingly, however, authoritarian rulers are turning to subtler strategies. Guriev and Treisman describe that when Peru's harsh censorship policies triggered sanctions and excoriation from the global community, President Alberto Fujimori changed tactics and paid large bribes to slant TV coverage.[90] When this failed with one station owner, Fujimori did not have him arrested or killed but rather stripped him of Peruvian citizenship—forcing the station owner (now a foreigner, so legally prohibited from owning Peruvian media) to forfeit his majority stake.[91] In today's backsliding Hungary, Orbán's success relies in part on his and his party's (Fidesz) ability to secure media influence through both cooption and direct ownership.[92] In the 2022 election, Orbán was reelected by a large margin after a campaign in which his opponent had been granted a total of *five minutes* of airtime. "It preserves the appearance of democracy," commentator Ira Glass observed, "After all, they're opposition parties. They exist. They can say whatever they want. They appear on television. But of course," Glass noted, "they don't get a real chance like they would in a real democracy."[93] As described earlier, authoritarians also rely on character assassination, libel, and defamation lawsuits in order to silence and discredit journalists, editors, and TV producers.

Although many observers predicted that the rise of digital media would undermine authoritarian control, smart authoritarians adapted with increasingly sophisticated methods. What Margaret Roberts calls "friction" involves "increasing the costs, either in time or money, of access or spread of information."[94] This might include "a slow webpage, a book removed from a library, reordered search results, or a blocked website," all of which raise search costs and thus deter access. Furthermore, Roberts describes the tactic of "flooding," or "information coordinated as distraction, propaganda, or confusion." Flooding, Roberts argues, "competes with information that authoritarian governments would like to hide by diluting it and distracting from it."[95] Yingdan Lu and Jennifer Pan find that official social media accounts of the Chinese Communist Party (CCP) circulate apolitical memes (e.g., funny pet or baby videos) to attract traffic to government sites (away from private ones).[96] Relatedly, Tony Zirui Yang describes how autocrats desensitize people to political censorship by censoring both political and nonpolitical information for purported social reasons (e.g., to protect the public from fraud or pornography)—to the extent that people support the practice.[97]

Contrary to the view that opening up civil society will weaken an authoritarian regime, the opposite has been true. People enjoy opportunities to engage in self-expression, cultivate their interests, and exchange ideas (about online gaming, sports, pets, birdwatching, and millions of other topics)—all of which improve their quality of life and create a churn of ideas. CSOs aimed at supporting needy communities—for example, the elderly, children, veterans, animals—provide services for which the government would otherwise be responsible. Furthermore, CSOs facilitate public deliberation and create communication channels between the people and the government. Siddharth Chandra and Nita Rudra note, "Select authoritarian regimes permit a fair amount of public deliberation, encouraging (controlled) expression of citizen preferences and often, actively soliciting them (e.g., surveys, workshops)."[98]

Indeed, scholars argue that civil society—a private media, CSOs, and public protests—provides a regime with what autocrats historically have lacked: information about the public's concerns and preferences, and about the competence and behavior of other officials.[99] Peter Lorentzen notes that protests convey valuable information because they "provide a clear division between groups whose grievances are tolerable and those with grievances severe enough to drive counter-regime challenges. Protests also serve as a useful device with which to monitor local governments, inhibiting corruption."[100] Furthermore, as a pseudodemocratic institution, civil society helps legitimate the regime in the eyes of the people and the international community.

NETWORKS, CONNECTIVITY, AND OPENNESS

Heterogeneity among authoritarian regimes includes significant variation in rights of mobility and openness. Scholars link a regime's engagement in international trade to selectorate size, arguing that whereas leaders with a small selectorate can pay off regime supporters through rent-seeking, authoritarians with a larger selectorate to satisfy "must look increasingly toward public goods such as free trade, whose effects fall on society widely. As a result, we expect that multi-party, and to a lesser extent single-party, autocracies will tend to prefer more open trade policies than non-party (often personalistic) dictatorships, monarchies, and military juntas."[101] Charles Hankla and Daniel Kuthy conclude that, with a larger selectorate to satisfy, and with longer time horizons encouraging leaders to foster policies that promote long-run economic growth, "more institution- alized types of authoritarian regimes will tend to adopt more open trade policies."[102]

In the late twentieth century, several developmental authoritarian coun- tries adopted growth models that depended on deep integration with the global economy. The United States provided its illiberal allies South Korea and Taiwan with market access, technology transfer, and support for their membership in multilateral trade and development institutions.[103] These and other authoritarian countries (e.g., Singapore and China after the 1980s) pursued a strategy of economic development reliant on market access, inte- gration in the globalized trade and financial regime, and (as detailed below) access to leading global educational institutions to raise the quality of their workforce.

Indeed, today's dictators permit their people to have greater mobility, both internally and abroad. Authoritarian regimes send students all over the world. In 2022 China sent over a million students abroad, and the top nine destinations were all liberal countries (of which the top three were Australia, the United Kingdom, and Japan).[104] In Vietnam's Doi Moi reforms in 1986, leaders "recognized Vietnam's need to be integrated into the world economy," which required "foreign trained personnel who understood the workings of the capitalist system." Vietnam initiated exchanges with Western universities and NGOs. Vietnam's openness and effort to raise human capital were welcomed and subsidized by the interna- tional community; "by 1995 Vietnam's educational sector had attracted nearly $40 million in bilateral ODA [Official Development Assistance], $20.5 million from United Nations agencies and $110 million from the World Bank and the Asian Development Bank."[105]

International mobility is not confined only to student exchange. Ivan Krastev writes that "most Russians today are freer than in any other period of their history. They can travel, they can freely surf the Web. . . . Unlike the Soviet Union, which was a self-contained society with closed borders, Rus- sia is an open economy with open borders."[106] Citizens of autocracies travel abroad in large numbers; "Russians made 45 million trips, Hungarians 26 million, Kazakhstanis 11 million, and Singaporeans 11 million."[107] Marlies Glasius agrees: "Nowadays, authoritarian states are much more likely to allow citizens to travel or migrate, but also return freely, and citizens of, for instance, Russia, China, or the Gulf states make ample use of these free- doms." Yet Glasius and other scholars note that "authoritarian regimes are seeing neither an unsustainable exodus nor a democratic transformation as a result of the increased mobility of their populations. Why not?"[108]

As they permit greater mobility, authoritarian leaders use various meth- ods to monitor and sometimes discipline their expats, and to isolate them from the public back home.[109] Tourists, businesspeople, and students overseas are often surveilled by consular and embassy staff or by co-nationals. For example, postindependence Algeria enjoyed significant economic gains from a large Algerian diaspora in France; to monitor them, "the Algerian state developed extensive mechanisms of surveillance and intimidation," introducing "*Amicales*, or 'friendship societies' linked to the Algerian Ministry of Interior, tasked with maintaining a close watch on migrants' activities."[110] In today's digital era, dissident political activity increasingly occurs in cyberspace; as Marcus Michaelsen argues, "Members of the Iranian diaspora, for instance, have used websites, blogs as well as satellite programs to provide audiences in Iran with alternative information and to participate in internal debates, circumventing state-controlled media."[111] Authoritarian regimes also use digital tools to monitor the activities of dissidents abroad.[112]

Surveillance sometimes leads to violence; between 2014 and 2021, Freedom House reported 735 cases in which governments targeted their nationals abroad "using attempted assassination, assault, deportation, and rendition." China was responsible for over 200 of those cases.[113] The most prominent case—a major departure from the practice of smart authoritarianism—was journalist Jamal Khashoggi's murder in Istanbul by agents of the Saudi Arabian government. The Saudi government engaged in a significant disinformation campaign framing Khashoggi as a terrorist and foreign agent.[114] A European human rights official noted the Saudi government's attempted public relations makeover: "For months the crown prince has been on a tour of world capitals to showcase the alleged transformation of his country into an open and tolerant society. He wants to portray Saudi Arabia as being at the forefront of progress in the Middle East."[115]

REFORMS

Although democratic advantage arguments expect that authoritarian leaders will struggle to reform—not wanting to disrupt a thicket of patronage relationships—scholars have shown that compared to democracies, autocracies are as able or better able to reform. Scholars highlight the role of crises and argue that leaders rely on "concentrated interests" in favor of reform.

Reformers frequently win support because serious crises make reform essential.[116] In nineteenth-century Japan, a broadly shared perception of crisis fueled support for the sweeping Meiji reforms.[117] In 1806 Prussia suffered a military disaster at Jena and punishing peace terms imposed at Tilsit. Reformists "saw in Jena and Tilsit the opportunity for putting their world back together in a new way"; the reform party "called for a dramatic reordering of Prussia's social, political, economic, and military structures."[118] Imperial Russia's disaster in Crimea prompted Alexander II's "Great Reforms."[119]

Indeed, political scientists argue that crises can convince people of the need for change. Notwithstanding arguments about the obstacle of concentrated interests, as John Waterbury argues, "the reform, liquidation, or privatization of SOEs [state-owned enterprises] is driven by fiscal crisis of varying intensity."[120] Ghanians reformed in the 1980s, argues Jeffrey Herbst, because "the economy had deteriorated to such an extent that even senior government officials, who normally benefit from access to imported goods even in times of shortage, reported that they were going hungry."[121] William Overholt agrees that, in some cases, "people are desperate; they fear total social collapse, and they often fear that families will not have enough to eat. Hence they are prepared to follow leadership that they find credible even at the cost of extraordinary social upheaval."[122] In 1980s South Korea, one of President Roh Tae-woo's advisors urged him to reform the *chaebol* conglomerates, warning that "any further delays in these reforms would invite revolution."[123] Economic crises thus have enabled leaders to push through otherwise politically challenging reforms.

The expectation of collective action theory that the "losers" of reform will block it is not borne out for a few reasons.[124] First, leaders can buy people off with concessions. In particular, Haggard notes that "the quiescence of labor in the face of wide-ranging reform" is a well-established phenomenon.[125] South Korea's Kim Dae-jung, for example, negotiated massive layoffs by making "important promises and concessions to labor."[126] Furthermore, collective action theory considers only the concentrated interests among the potential losers of reform—neglecting the interests of those who stand to benefit.[127] For example, South Korea's conglomerates supported financial reforms because "the privatization of the state-owned commercial banks enabled the *chaebol* to gain a toehold in an industry long closed to them." As a result of reforms, Chung-in Moon observes, "the *chaebol* moved aggressively into the newly opened industry."[128] Scholars also argue that collective action theory creates an artificial dichotomy of "winners" versus "losers," when—given the diversity of their business activities—many firms gain from reforms in some respects while suffering in others.[129] For all these reasons, political-economic reform is more feasible than collective action theory might expect, and helps explain the prevalence of successful reform.

Interrogating Smart Authoritarianism

Readers might have several questions in response to the discussion above. First, the model acknowledges a tension between economic growth and regime control; isn't it true, then, that democracy is a superior model because it does not face such tension? But democracies have their own challenges to manage.

As political theorists have studied for hundreds of years, democratic societies struggle to prevent the tyranny of the majority as well as the tyranny of the minority. Democracies must manage the need to address inequality against the tendency of wealth to accumulate in fewer and fewer hands—and the likelihood that the wealthy will use their wealth to acquire political influence to rewrite political rules. On the other hand, populist pressures risk undermining the foundations of economic growth and innovation. "Because all populists dislike checks and balances, the weakening of institutions (e.g., property rights protection, executive constraints, and contract enforcement by the independent judiciary) will result in lower investment, misallocation, and slower growth."[130]

Furthermore, many scholars argue that democracies cannot provide the conditions for the long-term strategic planning that is vital for economic development. First, democracies suffer from "short-termism."[131] Some types of innovation—notably science-based innovation—have long lead times before they yield results. But, in office for only short terms and always campaigning, democratic leaders "have relatively short political horizons" and "find their duties regularly interrupted by elections that distract from . . . long-term policy challenges." Additionally, focus on long-term growth is undermined because "politicians are rewarded for pandering to voters' immediate demands and desires";[132] investments in infrastructure and innovation "could potentially advantage future governments and hurt the government's own electoral chances in the short run."[133] Firms also suffer from short-termism: A desire to please shareholders leads firms to raise short-term profits and to return the majority of profits to shareholders, rather than investing in research and development.

Second, scholars also argue that special interests will interfere with strategic planning in democracies. Charles Lindblom argues that democracies are characterized by "the ease with which opponents of any positive policy to cope with a problem can obstruct it." He notes that democracies limit the power of the government by design, creating balancing institutions (e.g., the executive, the legislature, the judiciary). "To stop or block a change is a legal privilege granted to many."[134] Special interests also act as veto points. Lindblom notes that the private sector communicates with public officials via lobbying and campaign donations, often advocating policies that "conflict with the demands that the electorate makes"—that is, with the diffuse national interest.[135]

By contrast, scholars have argued that authoritarian leaders—with long time horizons and (relatively) shielded from special interests—can make and carry out long-term plans.[136] Authoritarianism, argues Jagdish Bhagwati, "shields the government from the rigorous and reactionary judgments of the electorate."[137] In the early Cold War, observers admired the command economy model for making "skillful use of the strengths of any large corporation: preparing and implementing long-term plans, using colossal financial resources for development in priority directions, carrying out major capital investments in a short period of time, spending large sums on scientific research and so on."[138] East Asia's developmental authoritarians (as well as leaders in Japan) used financial repression to compel household savings in order to mobilize capital.[139] Rather than letting market forces prevail, leaders identified what they deemed to be essential or promising sectors; they selected and supported "national champions" to which they funneled capital at low or even negative rates. A large literature argued that the East Asian experience defied the "Washington Consensus" about the advantages of liberal economic development.[140] Furthermore, when it comes to the ease of economic reform, scholars have argued that authoritarian leaders—relatively shielded from public opinion—could better implement the often-painful policies such as massive layoffs in sunset industries.[141]

Chapter 1 quoted L. Raphael Reif's vision of innovation as a wild garden, which suggests that innovation thrives if left alone. But perhaps the key word is *garden*, which by definition is not wild—perhaps plants do better if there's a gardener planning, watering, weeding, and keeping pests at bay. Indeed, many scholars advocate a strong government role in S&T policy. Warning that "America Could Lose the Tech Contest with China," Eric Schmidt and Yll Bajraktari lament "the absence of national technology priorities set by the government and a relative decline in government-funded R&D": They argue that American technology developed not in response to strategic goals but in response to "commercial interests."[142] Schmidt and Bajraktari argue that to compete with China, the White House must create and manage an American tech strategy. Such arguments reflect a revival of the evergreen debate about the role of the state in the economy, a debate that has reappeared in the United States in response to the rise of a statist competitor.[143]

It's also worth noting that the openness of democracies—so helpful for innovation—is seen by many as a source of vulnerability in their competition with illiberal societies: Information flows can be manipulated by domestic special interests and foreign adversaries.[144] Particularly in the age of social media, increased accessibility of media enables the spread of misinformation that reduces citizens' confidence in information and in government.[145]

In sum, arguments for a democratic advantage in economic growth typically compare the downsides of authoritarian regimes with the upsides of democratic ones. But a compelling argument about a democratic advantage cannot only weigh democracy's benefits against autocracy's costs. There is no theoretical reason to suggest that the inefficiencies inherent in an authoritarian country (particularly a smart authoritarian one) are going to be more economically harmful than the inefficiencies within democracies.

IS SMART AUTHORITARIANISM THE SAME AS
LIBERALIZATION?

Another critique of smart authoritarianism is that perhaps it just proves the institutions argument to be correct: that societies that are more liberal compete more effectively. Isn't smart authoritarianism just liberalization?

Smart authoritarianism is indeed associated with the adoption of more inclusive policies (e.g., financial liberalization, tariff reductions, greater central bank independence, low-intensity repression). Authoritarian regimes that pursue high-intensity repression and extractive economic policies remain economically backward—think North Korea, or Iran since the 1979 revolution. By contrast, regimes that adopt more inclusive policies can generate spectacular growth.

Regimes of this type, however, remain authoritarian. A smart authoritarian regime tactically and contingently frees up market forces, and allows greater social freedoms, for the purpose of encouraging growth and innovation. Lest this be confused with actual liberalization, note the experiences of Chinese tech entrepreneur Jack Ma or tennis player Peng Shuai. Both enjoyed freedoms for some time but lost them after the CCP began to view these individuals as threatening. Both people disappeared for a time, and then later surfaced occasionally.[146] Because freedoms are tactically and contingently granted (and withdrawn), such governments thus remain decidedly authoritarian.

GET SMART

The phenomenon of smart authoritarianism raises an obvious question: Under what conditions are dictators content to live by the old ways, versus when (like Kazakhstan's Nursultan Nazarbayev in 1991) do they invite Singapore's Lee Kuan Yew for a visit to chat about his smart authoritarian model? One explanation builds on a categorization from the authoritarian politics literature, which suggests that "personalist" regimes do not adopt the policies conducive to growth—in contrast to single- and multi-party authoritarian countries and military juntas, which more often do. Yet this begs the question of why a regime adopts policies that put it in the personalist category, versus reforming itself into a different, "smarter" one. As noted, Syngman Rhee's personalist regime presided over economic stagnation and instability in South Korea; after key reforms, Park Chung-hee's more institutionalized autocracy was an economic powerhouse. Chapter 4 details China's shift from Mao Zedong's personalism to Deng Xiaoping's more institutionalized CCP.

Scholars have already begun to explore the important question of what leads a regime to adopt (or discard) growth- and innovation-friendly policies. In a variant of the resource curse, scholars argue that countries with ample nature resources that provide an easy source of income (i.e., oil-producing countries) lack an incentive to pursue the painful reforms that foster diversified economic growth and innovation.[147] In the Soviet Union, for example, economic problems in the 1960s resulted in the Kosygin economic reforms (named after their sponsor, premier Alexei Kosygin), which many viewed as essential for reversing the USSR's economic slowdown. These reforms were abandoned due to the combination of soaring oil prices after the 1973 oil crisis and increased Soviet oil production in Western Siberia, which increased revenue to the government.[148] A subsequent moment at which reform was attempted—the reforms of the 1990s after Soviet collapse—also stalled as oil prices soared in the wake of Iraq's invasion of Kuwait. In Kazakhstan, scholars similarly argue that continued dependence on oil revenues will undermine economic reforms (and Nursultan Nazarbaev's vision of "Kazakhstan 2050," discussed in Chapter 5).[149]

Several scholars point to resource constraints plus a country's threat environment—the external environment, or combined threats from internal and external adversaries—as key explanatory variables that drive whether a regime chooses development over plunder. Richard Doner, Bryan Ritchie, and Dan Slater argue that developmental authoritarians arise in cases in which leaders are "staring down the barrels of three different guns," namely, the threat of "unmanageable mass unrest," a "heightened need for foreign exchange and war matériel induced by national insecurity," and "the hard budget constraints imposed by a scarcity of easy revenue sources."[150] They call this "systemic vulnerability," arguing that in such situations, leaders face incentives to supply not only private but also public goods.[151] In the cases of the East Asian "tigers," Mary Gallagher and Jonathan Hanson argue, "The political and economic constraints . . . increased the threat of revolution (in all likelihood combined with external invasion) whereas the choices for economic policy were limited by scarce natural resources and poor populations."[152]

Mark Zachary Taylor makes a similar argument about innovation, arguing that the policies required to encourage innovation necessitate important reforms that upset many sensitive domestic political relationships.[153] Governments will risk such policies, he argues, only when it's too dangerous to the regime (because of internal or external threats, or both) to *not* do so. Only in such circumstances will governments accept the costs and risks associated with what Joseph Schumpeter called "creative destruction."[154] In sum, while the conditions that have led some dictators to adopt smarter policies need to be more fully explored, scholars have argued that resource constraints and the threats facing dictators are key drivers of their policy choices.

The End of Dictator Drip

Students in my dictatorship seminar particularly admired Muammar Qaddafi's elaborate outfits, which featured capes, medals, maps of Africa, and pops of proud Libyan green. Indeed, *Vanity Fair* fangirled him as "simply the most unabashed dresser on the world stage."[155] Looking around the world today, my students wonder what has happened to "dictator drip"? Absent the Generation Z terminology, Guriev and Triesman argue that modern dictators have traded in their military uniforms and other regalia for conservative suits. No longer trying to shock or flaunt their differences with the liberal countries that hold the keys to economic opportunity, dictators—through their tailoring as well as their policies—are trying to blend in.

This chapter has presented the model of smart authoritarianism. Institutions theory expects that autocrats face a "king's dilemma" between exerting control (which will cause economic and technological backwardness) and loosening up (which will cause the regime to lose power). In this view, authoritarian regimes are homogenous and static, with extractive institutions that inevitably disadvantage them in a globalized world and information age.

The smart authoritarian view disputes these claims, arguing that authoritarian regimes are more heterogenous and adaptable than this view expects. In the smart authoritarian model, policies—not institutions—drive economic growth and innovation, and authoritarian regimes are highly diverse in the policies they adopt. By refraining from expropriation, creating a high-quality workforce, and responsibly managing their economies to attract capital, smart authoritarians create the conditions that encourage economic growth and even encourage innovation. And contrary to the expectations of institutions theorists, such policies actually enhance regime stability.[156]

The following chapter explores the mechanisms of smart authoritarian growth and innovation in the rise of China—whose economic, technological, and military rise not only defied skeptics but shifted the global balance of power.

China Gets Smart

"By the light of the burning palace which had been the pride and delight of her Emperors, she commenced to see that she had been asleep whilst all the world was up and doing," wrote Marquis Zeng Jize, "that she had been sleeping in the vacuous vortex of the storm of forces wildly whirling around her."[1] In his famous 1887 essay, Zeng lamented China's military defeats and humiliations at the hands of the great powers. The failure of China's Qing government to adapt to the changing technologies of the Industrial Revolution contributed to the Middle Kingdom's fall from regional dominance, and its descent into a failed state in the early twentieth century. After the founding of the People's Republic of China in 1949, the country remained poor under Mao Zedong's extractive rule.

But after Mao's death, a century after Zeng wrote his essay, China's government chose pragmatism over dogma, openness over autarky, and institutional constraints over authoritarian viciousness. The result was the most stunning economic rise the world has ever seen: China doubled its GDP every eight years, and 800 million people sweated and strived themselves out of poverty. China not only built the world's largest economy; Chapter 2 showed that in defiance of widespread skepticism, the country that had been dismissed as a copycat nation joined the ranks of the world's technological leaders.

This chapter analyzes modern China using the smart authoritarianism lens. The second section sets the stage by describing how China declined from being Asia's hegemon into the chaos of foreign invasion and civil war. Next, I describe how a set of more inclusive policies during the Mao Zedong era led to an essential prerequisite for economic takeoff: China's demographic transition. Moving into the reform era, the third section examines the mechanisms of smart authoritarianism that fueled China's catch-up growth. The Chinese Communist Party (CCP) created greater institutional constraints and adopted stable macroeconomic policies to encourage capital formation and foreign direct investment inflows. The CCP permitted a (government-controlled) civil society, opened up the economy to trade and investment, and shifted from high- to low-intensity repression. Chinese growth over time was accompanied by multiple rounds of reform. Finally, I address a counterargument to regime-based explanations for China's economic rise.

The Fall of China

At the start of the nineteenth century China had the largest GDP in the world (six times that of Great Britain), had historically been a technological leader, and was the center of Asia's politics, economics, and culture. But as the Industrial Revolution progressed, China experienced a dramatic decline.

Fearing social instability, China's Qing leaders failed to adopt the technologies that were transforming industry and warfare throughout Europe and North America. Leaders worried that steam power would lead displaced *junk* pilots to rise up against the regime;[2] the government saw railroads as "an explosive force too dangerous to handle" and as "the overthrow of time-honored custom and tradition, disturbance, and ruin."[3] China's growing technological backwardness led its GDP to fall, while the United States, Japan, and the European great powers experienced remarkable growth (see figure 4.1, below).

China's failure to adapt technologically led to military disasters. As Chapter 1 described, nineteenth-century battlefields showed an emerging gap between countries that were innovating with new technologies and those that were not.[4] China lost Hong Kong and Macao to Great Britain in the Opium Wars, and lost Taiwan, the Ryukyu Islands, and later Manchuria to Japan in the First Sino-Japanese War (1894). Other countries seized control over China's former spheres of influence, and demanded from Beijing steep war indemnities and privileges of extraterritoriality. A seven-country alliance of 19,000 troops invaded China in the wake of the Boxer Rebellion (1899–1901), eventually capturing and looting Beijing. Outcry over military humiliations fueled internal instability in China; the Qing were toppled and after 1927 the nation convulsed in a civil war between the Nationalists and the Communists. In the Sino-Japanese War that began a decade later (the war that Americans call World War II), the Chinese combatants paused in order to fight the Japanese. But after Japan's 1945 surrender, the Nationalists and the Communists resumed hostilities.

Out of the Chinese civil war emerged the People's Republic of China (PRC) led by Mao Zedong, who pledged to restore China to its former greatness. China had "stood up," Mao declared, and would attain equality with the superpowers. Mao's economic policies are widely pilloried, and indeed under his rule China remained mired in poverty. But the end of civil war and the founding of the PRC provided the centralization essential for

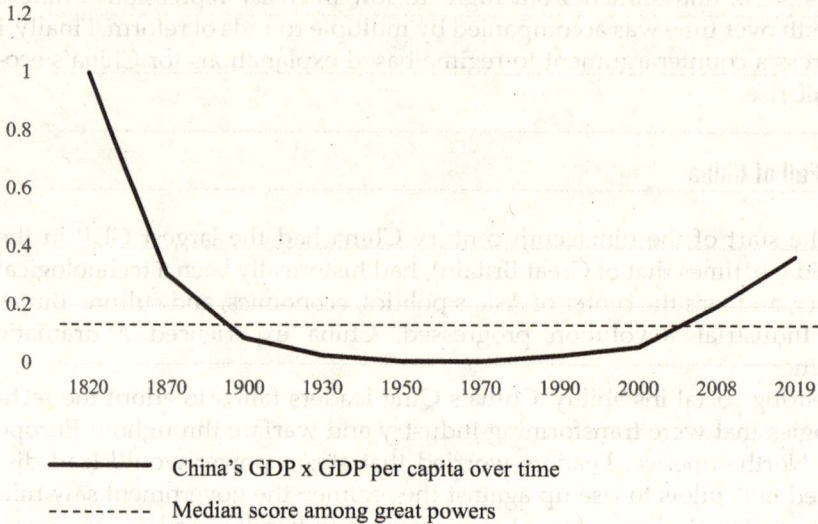

Fig. 4.1. China's GDP × GDP per capita over time.

economic growth, and some of Mao's reforms did lay important groundwork for China's later economic rise.

China's Demographic Transition under Mao

Mao Zedong is (justifiably) notorious for the tens of millions of people
killed in the famine of the Great Leap Forward and the terror and violence
of the Cultural Revolution. He is (justifiably) remembered for the draconian
control he imposed, and the poverty this created. However, less well known
is that several CCP reforms under his leadership triggered a demographic
transition: an essential foundation for China's later economic growth.[5] Policies related to women, marriage, and education were more inclusive relative to China's highly extractive past.

A "traditional society" cannot grow amid high mortality and high fertility.[6] Such conditions reduce the size of the labor force, because everyone
dies young and because women remain outside the formal workforce
engaged in childbearing and childcare. They are constantly pregnant, nurs-
ing, and caring for children—and then they die. High mortality and fertility
decrease savings and capital and foster low levels of productivity. For a
country to experience economic growth, mortality and fertility must fall.
This creates favorable conditions for growth because it enlarges the size of
the workforce (through longer lifespans and women joining the formal

workforce), promotes capital formation (through decreased consumption/ increased savings), and raises productivity (through higher parental investment in children's education).

China under Mao experienced a stunning drop in mortality. In fact, "China's growth in life expectancy between 1950 and 1980 ranks as among the most rapid sustained increases in documented global history."[7] In 1949, life expectancy at birth was about 35–40 years; by 1980 it had risen to 65.5 years.[8] Scholars attribute the drop in mortality to "the expansion of education during the 1950s together with large-scale public health campaigns," which explain 50–70 percent of the drop in infant mortality rates between 1960 and 1980.[9] Government public health campaigns during this era covered water sanitation, immunization campaigns, malaria prevention, and reproductive health. Mao's "barefoot doctors" famously improved public health in rural areas.[10]

In addition to the decrease in mortality, Mao's government enacted several policies—related to women, family, and marriage—that (eventually) reduced fertility. Before the PRC's founding, China had highly extractive norms and institutions regarding women and marriage, all of which elevated fertility: for example, child marriage, polygyny, concubinage, foot binding. Daughters were not counted when people counted their number of offspring; wives were property, bought and sold. "Women lacked all rights of property ownership and management and carried no formal independent decision-making authority in matters affecting the family and clan," writes Kay Ann Johnson.[11] Chinese women were "marginal members of the entire family system. They were temporary members or future deserters of their natal families and stranger-intruders in their new husbands' families."

Part of the communist revolution included changing the status of women in Chinese society. Kellee Tsai notes, "Both the Bolsheviks and the Chinese Communist Party (CCP) mobilized the most oppressed groups in society, namely peasants and women, to serve as the revolutionary base."[12] Under the slogan "Men and women are equal; everyone is worth his (or her) salt," the CCP in 1950 passed the "New Marriage Law."[13] The law required that marriage—previously considered a private affair—had to be legally registered with the government; it set minimum ages of 20 for males, 18 for women.[14] According to the law, marriage was "based on the free choice of partners, on monogamy, and equal rights for the sexes." The law sought to abolish practices—prevalent prior to the PRC's founding—in trafficking, parents forcing children into marriage, polygamy, and concubinage. The end of polygyny and child marriage—which frequently go together—was important for China's later drop in fertility. The eradication of foot binding—a brutal practice that crippled women and caused pain, infection, and numerous other health complications, including death among 10 percent of females—promoted women's health and extended female lifespans.[15]

Under Mao the CCP also expanded girls' access to education. Whereas in 1944 only 11 percent of girls attained a secondary education, by 1963 this rose to 54 percent.[16] Girls' literacy improved: Among girls born in 1940, 43 percent were literate, compared to 86 percent born in 1965.[17] Expanding girls' access to education delays childbearing and promotes contraceptive use, both of which reduce birth rates.

Despite these reforms, fertility in China initially remained high, at about six children per woman as late as 1970.[18] Mao's own attitudes toward population growth were erratic; the CCP had initially banned contraception (although it repealed this in 1954).[19] Mao also declared that childbearing was patriotic: that "strength in numbers" would help protect China in a dangerous world. "Under the Leadership of the Communist Party, as long as there are more people, miracles will be created!"[20] Mao's enthusiasm for large families, as well as improvements in medical and sanitary conditions, contributed to a baby boom after 1949. As one author recalls, in this pronatal environment, "My mother dutifully gave birth to five children. Our neighbour, Mrs. Wang, produced 11, and was declared a 'Heroine Mother' by the local authorities and given a large red rosette to pin to her lapel."[21] Thus, while changes in marriage and women's rights reduced fertility, the CCP was promoting pronatal norms.

Falling mortality amid sustained high fertility caused China's population to double from 500 million to 1 billion over the next 30 years. Increasingly alarmed about overpopulation, the government reversed course with antinatal policies. China developed a contraceptive pill by the mid-1960s and through the 1964 Birth Planning Commission increased efforts to distribute contraceptives and advocate family planning.[22] Posters, songs, and other forms of government propaganda encouraged smaller families and the use of contraception. Combined with the evolution in women's rights and girls' increased access to education, these campaigns were highly successful; by the 1970s, the percentage of women of reproductive age practicing birth control topped 80 percent.[23]

The government again raised the minimum marriage age for women (from 18 to 23) and for men (from 20 to 25), a practice that delays childbearing. By 1978, notes Kate Zhou, only 3.5 percent of women married before age 18.[24] In other words, it was not China's One Child Policy, enacted in 1979, that prompted the steep drop in fertility. "China's total fertility rate fell from close to six around 1970 to only 2.7–2.8 at the end of the decade. Thus, at least 70 per cent of the decline in fertility from 1970 up to the present was achieved prior to the launching of the one-child policy, not afterward."[25]

In the 1970s, Chinese policy experienced a "decisive shift" from optional to "mandatory and highly coercive family planning."[26] The State Council set ambitious population goals and oversaw a campaign described as "Later, Longer, and Fewer": later marriages, longer intervals between births, and fewer births. Rural families could have three children, and urban families could have two; families were closely monitored by local birth-planning workers and were penalized for exceeding this number.[27] Enforcement was often brutal. Bureaucrats harassed and bullied families, and often coerced people to undergo abortions and sterilizations. And of course, the highly extractive One Child Policy suppressed fertility even further, allowing one child per couple for urban families (two for rural and minority families).

Under the One Child Policy—adopted under Deng Xiaoping's leadership—the CCP swung away from more inclusive policies related to public education and family planning toward invasive government control. As many scholars have chronicled, the One Child Policy not only denied people choice about their families, it led to infanticide and many forms of abuse of women.[28] The policy created for China a super-sized "demographic dividend"—in the form of a massive labor force unencumbered by many dependents—that elevated economic growth. But as scholars have noted, the architects of the One Child Policy also created a massive problem—a super-sized "demographic penalty"—that will complicate future growth.

In sum, under Mao's leadership China remained poor due to numerous extractive policies (discussed below). At the same time, CCP reforms—related to public health, education, women, and marriage—caused a demographic transition that laid the foundation for economic growth.

Smart Authoritarianism and China's Rise

Since 1978, China's economy has grown at an average of 9 percent per year, sometimes achieving rates upwards of 13 percent.[29] Per capita GDP soared from $381 in 1978 to $12,175 in 2023.[30] While CCP data no doubt exaggerate China's growth,[31] China's economic development, and the effect on its people's lives, has been profound.

In contrast to the Mao era, growth in the reform years stemmed from higher productivity. From 1979 to 1994, Chinese productivity increased annually at almost 4 percent (compared with 1.1 percent during Mao's rule). This section argues that China's remarkable economic growth—built on the foundation of a demographic transition—was fueled by a shift from autarky, violence, and personalist politics under Mao to constraining institutions, low-intensity repression, and more inclusive economic policies.

CRISIS AND REFORM

China's dramatic political-economic reforms, which unfolded across multiple stages, contradict expectations that authoritarian regimes will be sclerotic—that is, unwilling to disrupt the complex patronage networks that keep leaders in power. To be sure, in China, "Economic reforms and market transition were profoundly threatening because they took away the monopoly over the distribution of patronage resources that political elites had enjoyed."[32] However (as described in Chapter 3), reforms were nonetheless possible. Scholars argue that China reformed in the wake of crises by buying off "losers" and by taking advantage of concentrated benefits among winners.

Mao's economic mismanagement resulted in a crisis so dire that reform was seen as the only option. From 1952 to 1978, China's centrally planned economy actually grew at a brisk 4–6 percent per year. This growth stemmed from improved education levels and increases in government investment; "Productivity actually regressed during this period."[33] Economic growth experienced significant ups and downs resulting from mismanagement and the paroxysms of Mao's campaigns. The rural collectivization associated with the Great Leap Forward of 1959–1961 created food shortages; the CCP's diversion of food from rural to urban areas caused 15 million–40 million farmers to starve to death.[34] Living standards had not risen despite decades of industrialization. GDP per capita was $175, and scarcity of consumer goods forced the CCP to impose rationing.[35]

People sought in a variety of ways to reduce draconian government control over their lives. As Kate Zhou describes, from 1977 to 1982 farmers "acted upon their revulsion with the government-controlled commune system, under which they had lived as de facto slaves of the government, by embarking on a massive exodus from the collective farms."[36] Some peasants tried to flee to Hong Kong. Before and especially during the Cultural Revolution, hundreds of thousands of Chinese "freedom swimmers" journeyed across 5–6 miles of open water, risking execution if they were caught by Chinese patrol boats, clinging to anything that would float, and braving shark attacks and currents sweeping them out to sea.

This desperation led CCP leaders to fear regime stability: "The traumatic experience of the Cultural Revolution had eroded popular trust in the moral and political virtue of the CCP."[37] After decades of chaos and deprivation, in the eyes of the Chinese people, notes Baogang Guo, "Revolution-based legitimacy was running out of steam."[38] Deng Xiaoping, jockeying for power in the aftermath of Mao's death, argued that the CCP had to reform or it would lose power. "How is socialism superior," he asked, "when our people have so many difficulties in their lives?"[39]

Pessimism about authoritarian reform is partly based on the argument that authoritarian leaders have less and poorer-quality information, but this was not the case in China. China has two kinds of media: public and internal. The public-facing media censor negative content about regime performance, in order to prevent mobilization against the regime. However, the internal media seeks to provide the Chinese leadership with the information it needs to address potentially destabilizing domestic problems.[40] Internal media informed Chinese leaders about the extent of the crises facing the country; Martin Dimitrov argues, "The top leadership was apprised of a number of negative phenomena, such as episodes of hunger; shortages of goods; and the incidence of corruption, theft, and waste."[41]

The crisis of the Cultural Revolution both weakened traditional opponents of reform and incentivized those opponents to support the effort. Susan Shirk notes that thousands of CCP officials had been transferred or banished to the country for "reeducation": this shake-up made "central party and state bureaucracies less daunting opponents to economic reform."[42] The bureaucracy was so weakened by purges (estimated by John King Fairbank to have taken out 60 percent of personnel) that officials had little power to resist reforms.[43] Barry Naughton notes, "The normal processes that rewarded officials and workers for compliance with the central government were in disarray, and the exchange of political 'goods' was disrupted."[44] He argues that civil servants thus favored an end to the turmoil, supporting Deng's efforts to create a stable career path within the CCP. Such efforts, Naughton notes, sought to "restore the political equilibrium in which patronage and promotion opportunities flow down the hierarchy in predictable patterns, in return for support for leaders running to the top."[45]

Quite usefully, crisis could be blamed on people distinct from current Party leadership. Deng's political machinations had led to the fall of the "Gang of Four"—die-hard conservatives who had sought to succeed Mao and continue his legacy. They served as convenient scapegoats for the disaster of the Cultural Revolution and the economic crisis facing China.[46] In sum, crisis helped propel reform, in part by reducing resistance from potential "losers."

"Reform and Opening Up" showed CCP officials the extent of the national crisis. In the 1970s, as the PRC began normalizing relations with other countries and Chinese officials began traveling overseas, "many of them for the first time, they were shocked and demoralized by what they saw."[47] They had expected that China lagged Europe and the United States; as Julian Gewirtz writes, Vice Premier Gu Mu traveled around Europe and "estimated that Chinese industry was at least twenty years behind Europe."[48] To their mortification, however, Chinese leaders discovered that they also trailed "even Japan and the newly industrializing countries of East Asia." The senior official Deng Liqun, after a visit to Japan, "reported with amazement that more than 95 percent of Japanese households had televisions, refrigerators, and washing machines, whereas almost none in China did."[49] As Orville Schell and John Delury write, "The shameful sense of living in a paradise lost, of having fallen so far behind other countries . . . would goad the country to strengthen and develop in order to finally catch up with the West and thus once again be able to defend itself and restore China to honor."[50]

Crises in China's foreign relations also encouraged reform. One of the newly industrializing countries outperforming China was Taiwan, then the Republic of China (RoC). The two countries were geopolitical competitors, each claiming to represent "China"; at the time the RoC held that privilege in the United Nations and other international institutions. Beijing regarded the RoC as a breakaway province and a core security challenge. In the 1970s, Taiwan's economy was outperforming the PRC's to an astonishing degree: with GDP growth rates at about 10 percent.[51] In 1980, Taiwan's per capita GDP was $2,370 (compared to China's at $307).[52] Furthermore, China's relations with the Soviet Union were worsening to the point of war. Although the USSR had been a key patron of the PRC in the 1950s, relations deteriorated to the point where the two nations "split" in the early 1960s. Later in that decade, the risk of war grew as their military forces skirmished on the Sino-Soviet border in 1969.[53] Thus both internal crises and external crises encouraged reform.

Over the years the reform process was far from smooth, featuring missteps and intense debate within the Party.[54] Reform featured many policy mistakes, notably an attempt at rapid price reforms in 1988 that ushered in a period of panic buying, runs on banks, and inflation.[55] The *fang-shou* cycle, notes David Shambaugh, is the constant push-pull between conservatives urging for control and modernizers encouraging reform.[56] Just as the smart authoritarian model prescribes, officials had to figure out how much freedom to grant people who were pushing for increased mobility, trading opportunities, entrepreneurship, and information.[57] Scholars note that some conservatives recognized the need for reforms, but insisted that the CCP could return to a perceived "golden age" of central planning before the onset of political chaos.[58] Rana Mitter notes the "fierce ideological battle within the leadership, reflected in society as a whole, about the downside of the reforms and the embrace of globalization." Conservatives opposed Special Economic Zones (SEZs) because these were growing richer, "while the factories of the impoverished west and north-east were unable to compete and having to shut down."[59]

Leaders reduced political backlash to reform by pursuing a gradualist approach rather than a "big bang" approach. Deng famously characterized his strategy as "crossing the river by feeling the stones"; he and Zhao Ziyang, the leading architect of reform, "were extremely cautious, 'taking one step forward and looking around before taking another.'"[60] Zhao commented that reform was "a process in which various interests and relations are readjusted and redistributed." One of his chief advisors noted, "We try to protect everyone's vested interests."[61]

For reforming the state-owned enterprises, for example, the CCP pursued a "dual track" system. Firms were required to produce products at a certain quantity sold at prices set by the state, but an alternative market track also existed; if firms satisfied their obligations to produce, they could produce beyond those quantities and sell at the (usually higher) market prices. "Managers and industrial bureaucrats," observes Shirk, "were protected by the security of the old system—the government continued to bail out enterprises operating in the red—while gaining access to the profitable opportunities of the market."[62] Because the market sector was experiencing strong economic growth, argues Naughton, this allowed China to "grow out of the plan" as it became a smaller and smaller fraction of the economy. In this way, Chinese leaders were able to push past authoritarian sclerosis and the opposition of concentrated interests, to oversee what has been called "reform without losers."[63]

HUMAN CAPITAL

A key dimension of reform was raising China's extremely low levels of human capital. Years of war had created a malnourished, illiterate populace; at the time of the PRC's founding, illiteracy rates stood at 85–90 percent.[64] During the civil war the Communists communicated with the people by using art, posters, opera, songs, and radio. Mao preferred to rely on such communication rather than raise literacy.[65] He and many other CCP leaders viewed education as a means to communicate ideology and prepare the masses for revolution. Other CCP leaders, however, saw education as essential to prepare workers for industrialization and collectivization.[66] The Party estimated in 1949 that 70 percent of its members were illiterate.[67] Bureaucrats tasked with imitating Soviet collectivization policies worried that illiterate local cadres would be unable to administer these policies, and so advocated education toward the goal of "creating a new class of peasants-turned-production-team-members."[68]

Mao did improve education in several ways (which, as mentioned earlier, contributed to China's demographic transition). To train a generation of teachers, the government adopted the Soviet education model.[69] Literacy rates rose; by 1959, the literate share of the population grew from 20 percent to 67 percent among people of ages 12–40.[70] The CCP also sought to unify the country linguistically under the common language (*Putonghua*) of Mandarin. Fairbank notes, "The Chinese writing system was not a convenient device lying ready at hand for every schoolboy to pick up and use as he prepared to meet life's problems. It was itself one of life's problems."[71] To promote the accessibility of Mandarin, the government adopted a romanized form (pinyin) in 1958, and in several rounds of reforms issued lists of simplified Chinese characters. In effect, language was an extractive institution that Mao's government reformed in a more inclusive direction.

In important ways, however, Mao neglected to develop and even stunted China's human capital. The country continued to lack universal primary and secondary education. Resources devoted to education remained tiny relative to both developed countries and other developing countries. Regulations required 10 percent of school curricula to be devoted to ideological and political study; actually, ideology wound through every subject.[72] Mao's campaigns—the Great Leap Forward and its resulting famine—reversed progress in rural education and contributed to significant rural/urban disparities.[73] During the Cultural Revolution, Mao swung Chinese education to the extreme ideological side of the spectrum, as he charged the people to find and drive out "representatives of the bourgeoisie who have sneaked into the Party, the government, the army, and various spheres of culture."[74] Schools closed in the initial turmoil, ceasing all admission of undergraduates for six years (graduate students for over a decade).[75] Mao shuttered many universities, creating polytechnic and teacher-training schools. Years of compulsory education were cut from 12 to 10. Policies during this time de-linked merit and reward: Entrance examinations were abolished, and students were chosen for university based on "redness" and "revolutionary" background.[76] The CCP took control of universities and abolished the fields of history, literature, and geography.

The terror and human cost of the Cultural Revolution were shocking. Scholars and scientists, and others who had been studying, researching, and working to improve their country's level of knowledge, were accused of harboring treasonous Western ideas. They were terrorized by Red Guards, crippled or killed by beatings or driven to suicide. People accused of ideological impurity could redeem themselves by beating up or killing others, often teachers or family members. Thus, while Mao's government made some strides toward improving Chinese education, his radical campaigns tragically undermined human capital and economic growth.

After Mao's death, China's smart authoritarian reformers embraced the functionalist view of education aimed at creating a workforce able to compete in a more complex, globalized economy.[77] The 1986 Compulsory Education Law made primary and secondary education free and universal, mandating nine years of schooling.[78] Private schools were permitted; the liberal arts returned to curricula. The competition that Mao abhorred was brought back, with political correctness taking a backseat to merit, requiring students to compete for university and graduate school admission based on entrance examinations. China's literacy rate rose from 65 percent in 1982, to 96 percent in 2018. In contrast to many developing countries, in which a variety of norms and institutions limit girls' education, China's literacy rates rose to over 99 percent among 15–24-year-olds, for both sexes.

Reform of higher education was a high priority. William Kirby notes that a key attribute of "world class" universities is that those institutions enjoy autonomy from the state.[79] As Chinese leaders sought to improve the quality of China's universities, they loosened the tight ideological control of the Mao era. Reforms passed in 1985 granted universities "more autonomy in enrollment, finance, and decision-making." Universities were allowed greater independence in "areas such as teaching, research, admission, international exchange and cooperation, management of facilities and finances, administration of faculty and students, and restructure of internal governance."[80] Party supervision of universities shifted from direct to more indirect methods of control. At the same time, universities still had to defer to the CCP. Reforms asserted the State Council's control over higher education, requiring it to operate within certain ideological bounds.[81] Ultimately, as Chapter 2 detailed, significant government investment in tertiary education led to dramatic increases in tertiary enrollments and helped fuel China's technological rise.

FROM PERSONALISM TO INSTITUTIONALIZATION

Scholars argue that the arbitrariness, unpredictability, and lack of leader constraint associated with personalist rule undermines economic growth, whereas greater institutionalization reduces uncertainty and supports economic development.[82] In China, after years of turmoil under Mao, Deng's efforts to institutionalize Chinese politics encouraged political stability and improved economic management.

Mao ruled as a personalist leader, creating a country in which there was no higher authority than Mao's words. Decisions were made on the basis of Mao's edicts, or on his associates' interpretations of his words and desires. "Depend on the rule of man," Mao declared, "not the rule of law."[83]

A vital aspect of Deng's reforms included a shift from personalist rule toward greater institutionalization, meritocracy, and predictability. To the CPC (Communist Party of China) Central Committee in 1978, Deng argued that China had to abandon revolution and personality cult to embrace the rule of law.[84] A Party directive laid out the goal of replacing "arbitrary, charismatic authority with regular, rule-based authority":

> In order to safeguard people's democracy, it is imperative that we strengthen the socialist legal system so that democracy is systematized and codified, and that these democratic institutions and laws possess stability, continuity, and great authority. We must have laws to follow; the laws that we have must be obeyed; enforcement of the laws must be strict; and violations of the law must be investigated.[85]

Toward ending the destabilizing purges and creating greater stability for Chinese officials, Deng oversaw important political reforms. He publicly pledged, notes Shirk, his commitment to accelerating the long-delayed process of political institutionalization"; crafting a system "governed by rules, clear lines of authority, and collective decision-making institutions to replace the overconcentration of power and patriarchal rule that had characterized China under Mao."[86] Naughton observes, "An enormous effort was made to rebuild and regularize the personnel system, recreating and strengthening predictable career incentives within the hierarchy."[87]

Term limits were an important dimension of such reforms. The party abolished lifetime tenure and instituted a mandatory retirement age. Pierre Landry argues that by combining top-down appointments with term limits, the CCP ensured that lower-level officials would answer to superiors, and that term limits would deny cadres the ability to establish corrupt fiefdoms that would make them more independent of central control. Term limits would also allow capable young officials to be promoted to replace older officials.[88] Notably, argues Xuezhi Guo, the 1982 constitutional reform represented "a major legal breakthrough for the PRC" that sought to prevent a concentration of power that could lead to the disasters of the Mao years.[89] The amended constitution sought to institute the rule of law, created term limits for the presidency and other political leaders, and created checks and balances by elevating the role of other actors in decision-making. Scholar Liu Songshan argues that by abolishing lifetime appointments for top cadres, the CCP implemented "a historic reform of China's political system and a fundamental change whose positive impact and progressive significance cannot be overstated."[90] Commented Andrew Nathan, the succession from Jiang Zemin to Hu Jintao "displays attributes of institutionalization unusual in the history of authoritarianism and unprecedented in the history of the CCP."[91]

Under Mao, civil servants earned their posts through "redness" rather than competence. Mao-era officials had emerged as leaders in the civil war and the war against Japan; they were "barely literate," having "not had any opportunity for university training during the chaotic war and revolution years of the 1930s and 1940s." For the civil service, Mao favored people from the peasantry or the working class—rejecting those from the "bad classes," the more educated people from the landlord and capitalist classes.[92]

Policies to encourage a more professional bureaucracy—though still a long way from a meritocracy—represented an important shift. Prioritizing economic development, the Party needed officials with skills to administer the economy, which required selecting officials based on competence rather than political loyalty or personal connections (*guanxi*).[93] Deng instituted new civil service guidelines in 1977, requiring examinations and rejecting class background as a criterion for selection.[94] Zhao implemented further reforms of the examination system in 1986. The more meritocratic nature of the civil service was echoed in education reform, where students sat examinations at every level of schooling; by the time a student was taking the civil service exam, then, that student had successfully demonstrated the ability to pass examinations since elementary school and all the way up.

Policy changes in the civil service also helped reduce corruption (though it remained prevalent). The 1990 Administrative Procedures Law empowered the Chinese people to sue local leaders; "it gave ordinary people the right to bring suit against rapacious, arbitrary officials."[95] At the local level, the introduction of village elections helped create leaders who could protect "villagers against the illegal predatory exactions of township and country officials, on whom they no longer needed to depend for their positions."[96]

Deng shifted the CCP from personalism in other ways. He and other Party leaders encouraged the greater influence of the "hitherto irrelevant" National People's Congress.[97] Starting in 1980, the CCP experimented with elections for local people's congresses. "Although all the candidates had to be vetted by the party, for the first time . . . the local population was given the opportunity to choose its own representatives."[98] Furthermore, Deng shared the stage with "a large number of important and influential revolutionary elders, who formed a kind of 'senate' of influential leaders. The most powerful were sometimes popularly referred to during the 1980s as the 'Eight Elders.'" Deng also elevated a number of younger officials, designating them as the next generation of leaders.[99] Some of these men were more conservative toward reforms, others more bold; the give-and-take that this created, Naughton argues, encouraged a gradualist approach to reform that served China well. It contrasts with the instability under Mao and with assumptions about authoritarian caprice.

Since the reform era, an extensive cadre training system has raised the quality of CCP governance. Scholars describe the vast system that features Party schools, national executive leadership academies, and colleges of administration, as well as universities. These teach ideological indoctrination as well as leadership and administration. The CCP "also spared no effort to sponsor overseas training for Chinese political elites to learn state-of-the-art expertise from advanced countries through first-hand observations and field trips," with over 700,000 officials sent overseas between 1978 and 2010.[100] In sum, efforts to reduce personalism and improve the quality of governance were important dimensions of the reform era.

PROPERTY RIGHTS

Although to some extent China's economic rise relied on improved property rights, China's dramatic economic growth occurred amid a weak legal system and the absence of a well-established property rights regime.[101]

During the Mao era, property rights did not exist. The Party seized people's possessions during campaigns against landlords and "cosmopolitans." In 1952, private firms comprised half of the Chinese economy, but by 1962 this fell to zero. During the Cultural Revolution, the legal system was "associated with capitalists, deemed inconsistent with class struggle, and utterly demolished."[102] The regime targeted landlords as enemies of the state; it confiscated agricultural land, forcing people to work on communes whose output was turned over to the government. "Usurping properties from the propertied class was a glorious act."[103] In the Great Leap Forward, grain production fell by 30 million tons from 1958 to 1959.[104] In the cities, businesses (small shops as well as large companies) were nationalized, and so the state sector was the only sector.

Under Deng, the CCP improved property rights and intellectual property rights in order to encourage private investment and foreign direct investment (FDI). In 1979 the National People's Congress passed its first law allowing FDI, and in the 1980s the government established SEZs. China's first patent law passed in 1985. China further strengthened its intellectual property (IP) regime when pressured by trading partners (e.g., in 1992, pressured by the United States) and by international institutions (in 2000, as part of China's ascension to the World Trade Organization [WTO]). Also in 2000, the Law on Solely Funded Enterprises took effect, guaranteeing the "legitimate property" of nonstate firms.[105] In 2007, a highly debated Law on Property Rights finally, "for the first time, explicitly places privately held assets on an equal footing with state and collective property."[106] Over time, Chinese intellectual property rights law has been "substantially amended to make it more aligned with those of many industrial nations. The strengthening of China's patent laws was also accelerated by its membership in the WTO."

Improvements in property and IP rights contributed to increases in FDI inflows.[107] Initial inflows came from Hong Kong, Macao, and Taiwan, via investors with linguistic, kinship, and other network ties to the mainland.[108] FDI eventually became global, ballooning from essentially nothing in 1979 to $45.5 billion in 1998, then to nearly $70 billion by 2006. This amount put China just beneath the United States as the second most popular destination for FDI.[109]

Although property rights protections increased with China's economic rise, an important point is that economic development occurred despite the fact that these rights remained highly underdeveloped. "Property rights (both cash flow and control over assets)," argue Jiahua Che and Yingyi Qian, "were fundamentally unprotected against state predation."[110] Argues Yue Hou, "Millions of Chinese private entrepreneurs face . . . heavy-handed government regulation [and] expropriation . . . on a daily basis."[111] Land ownership was a prominent issue.[112] Peasants suffer from "the sale of collective farmland out from under them by local cadres and officials," from which "the peasants receive only a small fraction of the proceeds from such sales, and other parties to the transaction benefit disproportionately."[113]

China's IP protections also remain underdeveloped. China's WTO obligations required it to enforce IP protections, as China's 1993 patent law appears to do; however, lack of enforcement has led to rampant IP theft and to vociferous protestation from foreign firms.[114] Foreign automobile manufacturers have taken Chinese firms to court. Chinese spies dug up genetically modified seeds from Iowa fields and shipped them to China; they burgled offices; and through cyber theft they stole vast amounts of proprietary information.

Rampant IP theft persists in China—for which the first Trump administration and then the Biden administration retaliated with punitive tariffs on Chinese exports.

The CCP understood the need to reassure foreign investors in order to attract capital, jobs, and technology transfer.[115] Confiscation was counterproductive, as it would deter firms from investing. As they sought to attract FDI, the Party tied cadre evaluations and promotions to meeting growth targets. Officials were encouraged

> to provide a hospitable environment for factories, typically through the provision of local, public goods such as establishing a basis for secure rights of factory owners, the provision of infrastructure, utilities, access to markets, and so on. Those jurisdictions that fail to provide these goods find that factories move to other jurisdictions. Local economic activity and tax revenue decline as a consequence.[116]

Local officials "could exercise property rights directly or bestow them on a specific set of investors within their own jurisdictions."[117] Scholars found that within China, variation across time and space shows the importance of property rights as a driver of FDI: FDI increased with the strengthening of the property rights regime, and regions that better protected property rights were more successful in attracting FDI.[118]

While property rights and IP rights did improve in China over time, many experts have observed that the country did not develop as neoliberal theory would have expected. In contrast to generally accepted theories about formal property rights and the rule of law, Chinese officials and the private sector each devised "functional substitutes for the formal institutions that receive so much attention in the analysis of developed countries."[119] One such "functional substitute," argue Che and Qian, came in the form of firms owned by local governments (Town and Village Enterprises, or TVEs), which came to dominate China's private sector. Che and Qian argue that TVEs "emerged as an organizational response to imperfect state institutions, which may work better than either conventional private or state ownership." While imperfect, "this type of ownership can reduce the adverse effects of state predation in the absence of institutions to constrain the state."[120] China experts have also examined how entrepreneurs and firms have behaved in order to minimize their risk of expropriation by the government. Scholars have identified a "toolbox of strategies," including fostering political connections, working with business associations, and pursuing CCP membership and legislative office.[121]

In sum, the CCP encouraged economic growth through a process of what Yuen Yuen Ang calls "directed improvisation": a co-evolution of markets, property rights, and development strategies.[122] The bottom line is that China's smart authoritarianism led it to provide (albeit imperfectly and in ways not recognizable to neoliberal economic theory) protection against confiscation that encouraged private sector investment and FDI.

PRIVATE GOODS

Patronage, rent-seeking, and other kinds of corruption are common means by which authoritarian leaders reward supporters.[123] A prominent argument holds that corruption undermines economic growth and innovation. Yet the case of China stands as a paradox of high growth amid high corruption.[124] China's rise has featured corruption of a scale that Arthur Kroeber calls "extraordinary . . . routinely running into the hundreds of millions if not billions of dollars."[125] Corruption indices rank China as #87 out of 180 countries, tied with Serbia; most advanced industrialized economies have far lower rankings in terms of the prevalence of corruption in their economies. For example, Denmark and New Zealand are #1 and #2; Japan ranks at #18, and the United States at #22.[126] How was such a corrupt country able to experience such dramatic economic growth?

Scholars have argued that the CCP (like previous "developmental authoritarians") experienced economic growth amid high corruption because of the nature of corruption. First, the postreform CCP, like other smart authoritarians, discouraged the worst forms of elite predation, which raise inflation and overvalue the currency. Predatory regimes in Venezuela, Nigeria, and Zaire, for example, relied on currency overvaluation and import licenses, which devastate the economy. Chinese leaders in the reform era avoided such behavior. Indeed, notable reforms included rounds of devaluing the previously overvalued yuan (1981–1985).

Second, Ang argues, understanding Chinese growth requires "disaggregating" the corruption there. Noting different types of corruption, she argues that "petty theft" and "grand theft" are "the most economically damaging as they drain public and private wealth." This kind of corruption, she writes, "subverts law and order, deterring investors, tourists, and even foreign aid donors."[127] Such corruption is rarer in China, by contrast to what she calls "access money." From a firm's standpoint, Ang argues,

> Access money is less a tax than an *investment*. For example, Chinese entrepreneurs are willing to bribe their way into legislative congresses because the benefits of networking with Party-state bosses more than offset the expense. Likewise, in the United States, big corporations sink billions of dollars into lobbying every year because returns exceed costs. By enriching capitalists who pay for privileges and by rewarding politicians who serve capitalist interests, access money can perversely stimulate commercial transactions and investment, which translates into GDP growth.[128]

Ang finds that other countries in which access money is the dominant form of corruption include the developmental authoritarians (e.g., Singapore, South Korea, Taiwan), as well as the United States. Other scholars have proposed similar explanations for the paradox of Chinese growth amid high corruption. Chong-en Bai, Chang-Tai Hsieh, and Zheng Song argue, "Chinese private firms succeed, in part, by obtaining a special deal from a local political leader, which enables them to either break the formal rules or obtain favorable access to resources."[129] For example, under Deng's reforms, valuable rents included low-interest government loans to finance projects, and opportunities for firms operating in the SEZs.[130]

Most importantly (as was also the case in South Korea and Taiwan), opportunities to receive valuable rents were linked to performance. Government officials who successfully oversaw economic growth in their jurisdictions were given large cash bonuses and promoted.[131] "Reform policy entrepreneurship," notes Naughton, was rewarded "by being designated an 'experiment' or a special zone," which opened up access to resources.[132]

Scholars have argued that while perhaps China's economy can grow for a while amid high corruption, such growth would invariably fall apart.[133] Xi Jinping, however, unleashed a sweeping campaign against corruption. In 2011, the CCP's Central Commission for Discipline Inspection "conducted formal investigations into 137,859 cases that resulted in disciplinary actions or legal convictions against party officials."[134] The following year, Xi Jinping proclaimed that corruption posed an existential threat to the CCP, calling the extent of corruption "grave" and "shocking."[135] In 2018, Xi launched an anticorruption campaign with the slogan "Sweep away black and eliminate evil." This was hailed as a great success: "Nearly 40,000 supposed criminal cells and corrupt companies busted, and more than 50,000 Communist Party and government officials punished for abetting them, according to official statistics."[136] Although some observers note that Xi's campaign was motivated by an effort to purge political rivals, corruption is indeed being curbed.[137]

HIGH- TO LOW-INTENSITY REPRESSION

Political scientists argue that high levels of repression undermine economic development. Previously in China, high levels of terror and violence repressed economic growth, in part by isolating China from the international community. Now, economic development could take off, in part because such repression was eased as part of reform.

Mao Zedong ruled through terror and violence, instituting campaigns to weed out rivals and deter potential enemies. In Mao's purges and campaigns, writes Frank Dikötter, "thousands were locked up, investigated, tortured, purged, and occasionally executed." After the founding of the PRC, land reform forced villagers "to beat and dispossess their own leaders in collective denunciation meetings, accusing them of being 'landlords,' 'tyrants,' and 'traitors.' Some did so with relish, but many had no choice as they risked being targeted themselves. Close to two million people were physically liquidated," their possessions seized by the Party.[138] The violence of land reform was followed by Mao's Great Terror, which smashed Chinese civil society, purged the civil service and business communities, and burned books in order to discipline artists and writers.[139] Mao next used the violence of the Great Leap Forward—which involved dispossession and collectivization of properties and businesses—to weed out enemies. The Great Leap Forward killed upwards of 45 million people through famine and violence.

The resulting economic disaster put Mao on the defensive, leading him to lash out with more terror. Some Party officials were calling for a new direction. Mao's paranoia surged in 1956 when Nikita Khrushchev denounced Josef Stalin and his cult of personality. Feeling vulnerable to such a movement at home, Mao fought back. Calling for dialogue in the "Hundred Flowers Campaign," Mao was stunned by the criticism that poured forth. He responded with a political purge and another mass terror campaign: the Cultural Revolution. "Anyone who had expressed reservations about the Great Leap Forward was hunted down, as some 3.6 million party members were purged as 'rightists.'"[140] In the Cultural Revolution the Red Guards, acting at the behest of the CCP, committed terrible violence against hundreds of thousands or even millions of people. People were beaten, tortured, executed, and driven to suicide; others were sentenced to prison, to forced labor, or to internal exile.[141] Revolutionary committees consisting of military and civilian authorities conducted the "Cleansing the Class Ranks" and other violent campaigns. Andrew Walder writes, "Those who were targeted in these campaigns suffered some combination of investigation, imprisonment, coercive interrogation, expulsion from homes or jobs, or a political verdict that stigmatized them and their immediate family members for years thereafter." [142]The Mao era was thus associated with high-intensity repression that tormented and shattered the lives of tens of millions of Chinese people.

Under Deng, repression initially remained high. In the early 1980s, Deng advocated responding to a crime wave with highly publicized, campaign-style methods dubbed "strike hard" (1983–1986). The CCP "dramatically accelerated the procedure for trying cases, approving verdicts and executing sentences" for violent crimes; the campaign featured mass arrests and high execution rates.[143] The 1989 Tiananmen Square protests and mass killings then proved a turning point, after which the CCP adopted new public security tactics.

In the years after the Tiananmen crisis, the CCP shifted to less overt and less brutal methods of control. Ang writes, "The ruling party has developed a much more sophisticated arsenal of strategies for repression. It knows that a violent crackdown in public view is too costly."[144] As the CCP shifted to a model of more selective repression that was more compatible with economic development, scholars noticed a "considerable fall in the level of overt political repression."[145] The party's strategy, argues Eric Li, "focuses on containing a small number of individuals who have political agendas and want to topple the one-party system." For example, the government does not murder and torture journalists but rather deters them from publishing unwanted coverage through cooption, house arrest, and threats of imprisonment (often for libel, defamation, or nonpolitical crimes such as tax evasion). As of 2022, China had only 50 journalists in detention.[146]

The CCP reformed its internal security practices and apparatus. The CCP "incentivized local authorities to respond aggressively to citizen protests (whether through repression or concession), notably by making internal security a key feature of cadre performance evaluation. The CCP also elevated the bureaucratic rank and status of security chiefs, signaling the great importance it attached to public security.[147] The internal security apparatus expanded its numbers of police and informers.[148] Yet repression grew more focused and less violent. The CCP "reworked the Party political-legal apparatus to address citizen grievances in a more flexible and coordinated manner."[149] The central government "urged lower-level officials to eschew the use of force whenever possible. . . . And local cadres who turn to force too quickly (or unsuccessfully) are often punished with lost bonuses, demotions, or even imprisonment."[150]

During the reform era, political protest grew as reforms created unemployment and dislocation. "State-owned enterprises began massive layoffs, the private sector grew dramatically, and many farmers left their land to pursue work as construction or factory workers in the urban areas, creating a 'floating population' that had reached an estimated 79 million."[151] Citizens protested against the growing corruption of local authorities, who imposed taxes and fines, and whom the CCP found increasingly difficult to monitor.[152] Thus while reform benefited the economy overall, many people were adversely affected, and they expressed their discontent through protest.

As protests increased, the CCP created boundaries for the types of protests it would tolerate. Yao Li finds that the CCP "has become more willing to tolerate protests addressing economic, environmental and even anti-corruption issues."[153] Analyzing protests from 2001 to 2012, Li finds that "authorities tolerated the majority of protest events and that most protest events did not take transgressive forms—staying away from violence, from radical political claims (for example, opposing Communist rule), and from linking organizationally with other protests."[154] Similarly, Peter Lorentzen finds that the CCP tolerates what he calls "loyalist" protests: those seeking "benevolent treatment from a higher level of government." Citizens usually have respected the bounds demanded by the regime: keeping protests narrow (that is, not seeking to connect to other movements), keeping to "local grievances and material demands," and using loyal rhetoric that looks to the CCP rather than challenging "the CCP's right to rule or the overall direction of its policies."[155] Scholars argue that this approach has served the regime well, actually enhancing its stability. Protests play an important role in alerting the regime about local mismanagement and corruption, allowing "the regime to identify and deal with discontented communities before they turn to more extreme counter-regime activities or revolt."[156]

By contrast, the CCP quashed what it deemed "transgressive" protests. Notes Yao Li, "Claims for democracy and efforts to contest the Communist Party's monopoly on political power are prohibited or strictly constrained." Hence the crackdown at Tiananmen Square (and later in Hong Kong), which challenged CCP authority. Furthermore, the Falun Gong spiritual movement connected people from all over the country, and showed itself as "the type of well-run, mobilized, nationwide group the government fears." The CCP targeted the Falun Gong with brutality. "Chinese authorities reportedly have executed several Falun Gong adherents, locked up hundreds in psychiatric hospitals, and imprisoned thousands of others."[157] Because protests from ethnic minorities—such as Tibetans and Uyghurs in Xinjiang—create the possibility for transnational mobilization and international interference, the CCP cracked down hard in those areas as well. Toward the Uyghurs, for example, the CCP switched to a noticeably more punitive policy—more intense and more collectivized repression of the Uyghur community—as the regime perceived a higher risk of "emerging contacts between Uyghurs and Islamic militant organizations in Southeast Asia and the Middle East." The CCP acted "to block diffusion of terrorism into China and preventively inoculate its population from infection by extremist and terrorist networks."[158]

Furthermore, defying the view that authoritarian regimes cannot survive increased information flows, the CCP adapted public security to the digital era. The Ministry of Public Security "firmly established the primary importance of public security informatization at the Nanjing Conference in 2008." In its embrace of public security informatization, argues Peter Mattis, the CCP shifted its measures "from reactive to preemptive through the use of intelligence collection and synthesis."[159]

Aided by increased intelligence about protest movements—which grew exponentially in the digital era—the CCP shifted toward preemptive and targeted methods of repression. Kevin O'Brien and Yanhua Deng point out that in addition to traditional policing, "judges and court staff may be sent to the streets to buy off demonstrators, housing officials may be empowered to give rural evictees the right to move to cities, and retrievers may be paid bounties to surveil and intercept persistent petitioners to ensure that they do not make it to Beijing." O'Brien and Deng conclude, "Whether they rely on money, bargaining, or coercion, one common feature of these approaches is that they are directed at individuals and are designed to get a person off the street not only today but also in the future." The CCP has adopted what O'Brien and Deng call "relational repression": the regime puts pressure on relatives (by threatening their housing or career prospects) to pressure their loved ones to cease political activity. Such activity shifts repression from the public sphere (with all its attendant risks of backlash and global opprobrium) to the private: It "turns repression into a highly charged conversation with family members, neighbors, or old friends, and uses people who have a hold over the protester to deliver the state's message to desist."[160]

Furthermore, Rory Truex explains that the CCP monitors a "dissident calendar," anticipating "historical anniversaries, high-level regime meetings, and even international sporting events" in order to preempt dissident activity. Truex argues that the regime relies on a "'catch and release' dynamic that allows the regime to get through the sensitive dates that might produce broader, overt collective action. Dissidents are taken off the streets, intimidated and mined for information, and released once the danger has passed. Attention-grabbing trials are often avoided entirely, minimizing the risk of popular backlash."[161] In sum, over time the CCP has transformed its methods of control from the overt and brutal methods of the Mao years to more targeted, low-intensity repression.

FREEDOMS AND CIVIL SOCIETY

The CCP has also followed the smart authoritarian playbook by allowing (government-controlled) civil society. Scholars note that civil society allows for public expression and cultivation of interests; it aids in policymaking and reform by encouraging communication of grievances and by flagging areas of government failures. Civil society in China has enabled the CCP to sustain and strengthen Party control.

The CCP's acceptance of civil society came in the reform era, after more than 20 years of the state controlling every aspect of people's lives. "Under the leadership of Mao Zedong, programs such as household registration, job allocation, distribution of food coupons, and the commune system served to restrict people's mobility and control their daily life. Meanwhile, official education, media, arts, and literature ensured ideological uniformity."[162] After the trauma of the Cultural Revolution, however, people pushed for more rights: The recreation of civil society, argues Zhou, was a "movement undertaken by individual citizens who were fed up with the government controlling every aspect of their lives": they wanted "more non-governmental space and more individual rights."[163]

Economic growth, urbanization, and other aspects of China's transformation also created tremendous churn within society that sought expression and government attention. "Demands on the leadership are emerging from new proprietors such as private businesses and homeowners, citizens such as environmentalists who feel they have a personal stake in China's future, and the newly needy, such as migrant laborers and poor formers," wrote George Gilboy and Benjamin Read. "Despite repression, these communities are experimenting with ways to organize and act on their interests."[164] The Party, Yanqi Tong argues, recognized that without approved methods of expression to articulate their interests and grievances, "people would have to employ abnormal and extreme means to articulate their demands, such as riots, demonstrations, and strikes."[165]

The CCP allowed a significant expansion of civil society. The CCP welcomed the private business sector (the "Red capitalists") as Party members.[166] Furthermore, the number of registered civil society organizations (CSOs) soared to an estimated one million.[167] They "have been allowed or have created an increased organizational sphere and social space in which to operate and to represent social interests, and to convey those interests into the policy-making process. They not only liaise between state and society but also fulfil vital welfare functions that would otherwise go unserved."[168] China's civil society, argues Shaun Shieh, "has been an important source of bottom-up ideas and models that have developed independently of the Party-state since the early 1980s."[169]

Access to information also increased, particularly as the Party permitted the rise of private media. In 1979, only 69 newspapers existed, all state-run: "glorifying local and national leaders on the front page, and invariably positive reports written in formulaic, ideological prose inside."[170] In the reform era, newspapers were weaned off subsidies and allowed to sell advertising. The media, wrote Andrew Nathan, became "more commercialized and therefore less politicized": "They often push the envelope of what the regime considers off-limits by investigating stories about local corruption and abuses of power."[171] By 2005, China published more than 2,000 newspapers and 9,000 magazines.[172] As part of "reform and opening up," the CCP allowed the people to access foreign media such as the BBC and VOA.[173] Later, China would develop an exploding social media market—the world's largest, with 829 million users.[174] Its three social media giants are known as "BAT": Baidu, Alibaba, and Tencent. Shirk notes that leaders are ambivalent about private media; it provides "more accurate information regarding the preferences of the public to policymakers," yet it risks airing antiregime views. Shirk adds that an active media also provides top authorities with information about local governance; national leaders "use media as a watchdog . . . so they can identify and try to fix problems before they provoke popular unrest."[175] The CCP's internal media (as described earlier) continues to play this role as well.

Although Chinese civil society has grown significantly, the Party has retained control. In the Mao era, the 1954 constitution guaranteed freedom of religious belief, but people practicing religion were among those persecuted during the Cultural Revolution. The reform era allowed the return of religious practice under the 1982 constitution (restricted to five government-approved religions). Religious organizations are required to register with the government, and adherents must worship in those state-affiliated religious institutions—in which leaders have been vetted and co-opted by the CCP. According to Freedom House, "Government officials see religion as a tool of the party and vow to use secret agents to infiltrate and 'quietly smash' any religious groups operating outside of state control," for example, in underground churches.[176]

The flourishing of CSOs has not led to liberalization, but to the opposite. Argues Timothy Hildebrandt, CSOs "are granted enough space to meet their own, often narrowly defined goals, but not so much autonomy that they might challenge the government or otherwise undercut state interests. Social organizations work to further their own goals; at the same time, they often work to assist the government in implementing its policies."[177] Tony Saich notes that CSOs learned that to succeed they had to defer to the CCP.[178]

Furthermore, the CCP controls the flow of information through a variety of strategies. The "Great Firewall" blocks Facebook, Twitter/X, and YouTube; Google and LinkedIn withdrew or ceased social media operations after years of trying to operate in a censored environment. The CCP harasses and often expels foreign journalists. A massive internal security apparatus monitors and censors online content. The CCP communicates weekly censorship guidelines to editors and producers; private firms tend to employ their own censors so they will not run afoul of the government. Learning the type of coverage permitted by the CCP, journalists self-censor. As a result, "Financial publications like *Caijing* and *Southern Weekend* boldly evaluate companies' performance and expose corruption, though they rarely delve into political or social issues."[179] Journalists who remain within approved bounds enjoy career advancement; those who venture into "no-go" zones are fired or demoted, hit with libel lawsuits, or imprisoned. "The seemingly chatty, freewheeling press is not really freewheeling at all. The Chinese Communist Party is just more cunning about how it controls public opinion."[180] For all these reasons, in 2024 the group Reporters Without Borders ranked China 172 out of 180 countries.[181]

Methods of CCP information control have grown increasingly sophisticated. Censorship and disinformation are calibrated in anticipation of, and in response to, political events.[182]

Margaret Roberts argues that the CCP controls information not only through censorship but through tactics of "friction" and "flooding." The former raises search costs (e.g., by slowing the speeds of websites or "misplacing" books or documents). Flooding, by contrast, is the practice of churning out information (including apolitical information) in order to compete with, dilute, or distract from political information. Roberts concludes that even though censorship is quite porous, these other tactics have been highly successful for the CCP. "Only a minority of citizens who are interested enough in the information and have the education and resources to pay the costs of evasion are motivated and equipped enough to circumvent censorship." Most people, she notes, "are not willing to spend significant time becoming informed."[183]

When people think of authoritarian propaganda, they probably don't think of cat memes. But the CCP has responded cannily to the rise of digital media. Journalists, bloggers, netizens, publishers, and many others may experience intimidation, threats, or arrest. The government employs legions of fake accounts to spread disinformation and noise. One form of flooding is the CCP's increasing use of nonpolitical content, as Yingdan Lu and Jennifer Pan have demonstrated. On social media platforms, government accounts rely not only on propaganda but also on "clickbait" content—for example, listicles and memes—to drive traffic to state-run accounts.[184]

So far the CCP has been able to enjoy the benefits of a private media while keeping it under control. Indeed, many Chinese people do not feel the presence of censors, and censorship over time has become normalized.[185] Thus, in contradiction to the expectation that the expansion of civil society encourages political liberalization, the CCP has used government-controlled civil society as a tool to increase regime legitimacy and durability.

OPENING UP

One of the most famous photos in US-China relations is tiny Deng Xiaoping donning a massive cowboy hat at a Texas rodeo. The photo is memorable not only for its juxtaposition: a Communist leader embracing a cherished Texas symbol. The photo also startles because only a few years before, during the Cultural Revolution, a Chinese person donning a quintessentially American hat would have been beaten by a mob.

Mao Zedong had pursued a policy that Richard Nixon described as "angry isolation."[186] This isolation started with the Chairman himself; Jonathan Spence writes that "with the world outside China, Mao had virtually no contact."[187] The only country Mao had ever visited was the Soviet Union; he went twice during Stalin's rule. But he never returned after Stalin's death because of his anger at Khrushchev. Given China's vastness, "Why tour the four continents?" Mao declared in 1958.[188]

CCP policies kept the rest of China isolated from the world. During the Korean War the Party launched an antiforeign campaign, in which all foreign business assets were frozen and almost all foreigners fled the country.[189] In the 1950s, Mao had cultivated a close partnership with the Soviet Union; Soviet equipment and expertise (thousands of advisors) played an important economic role in the PRC's early years. But with the Sino-Soviet split—and the withdrawal of Soviet economic and military aid—China lost its key patron. With "no alternative technology partners and very little market access to technology," Mao led China into a state of "technology autarky."[190]

Mao deemed reliance on other states "dangerous," arguing that growth comes from "doing the utmost for itself as a means toward self-reliance for new growth, working independently to the greatest possible extent," and "making a principle out of not relying on others."[191] This strategy was in essence an import substitution strategy.[192] During the Cultural Revolution—an anti-cosmopolitan, anti-Western ideological campaign—"China broke diplomatic relations with three countries, strained its relationships with over half a dozen more, recalled all but one of its ambassadors, and cut down substantially its import of industrial plants and equipment from foreign countries." Mao, summarized Frederich Wu, pursued self-reliance "to the extreme of autarkical self-isolation."[193]

Deng Xiaoping's "Reform and Opening Up" thus marked a dramatic break and was instrumental to fueling China's catch-up growth. Under the "Four Modernizations," economic interdependence was no longer seen as exposing China to predatory imperialists, and was framed as compatible with Maoist thought (in contrast to the Gang of Four, who were blamed as having warped true Maoism.) Noted Wu, "The new leadership now regards foreign trade as a key element of China's Four Modernizations. The previously sensitive policy of relying on the import of up-to-date scientific, technical and managerial know-how and capital equipment to spur China's modernization is no longer taboo."[194] Hua Guofeng endorsed an outward-oriented economic model at his 1979 speech to the National People's Congress:

> Economic exchanges between countries and the import of technology are indispensable, major means by which countries develop their economy and technology. It is all the more necessary for developing countries to import advanced technology . . . in order to catch up with those economically developed. . . . We hold that the development of economic, technological, scientific and cultural exchanges and cooperation among various countries on the basis of equality and mutual benefit will help to promote their friendly relations and preserve world peace.[195]

As foreign trade expanded, the CCP reduced tariffs to levels comparable with the leading Organization for Economic Cooperation and Development (OECD) economies. China's ascension to international institutions helped significantly in this process: "The reforms associated with the evolution of Special Economic Zones and joining the WTO have all greatly reduced the potential for rent seeking from external trade barriers."[196]

Furthermore, financial reform in 1977 ended China's "long-standing reluctance to accept long-term loans from foreign sources,"[197] and, as noted earlier, joint ventures with foreign firms were legalized in 1979. The country whose leaders once gave shrill speeches about fighting exploitation by the evil capitalists now tried to lure foreign investors with incentives and preferential terms to set up shop in order to take advantage of cheap Chinese labor. FDI to China—as noted above—soared.

The Four Modernizations led to an explosion of both inward and outward tourism. Cruise ships flowed into Chinese ports after 1978, and the government was deluged by hundreds of thousands of visa applications. In 1978, China welcomed 124,000 visitors, equivalent to the number of people who had visited China over the previous 24 years.[198] The CCP reversed course on its Cultural Revolution–era hostility toward foreigners, instead publishing maxims "preaching friendliness, service, and courtesy toward 'foreign friends.'"[199] Language schools sprung up to train tour guides and other members of the rapidly expanding tourism industry. "Reform and opening up" also permitted Chinese people to travel overseas. After 1983, tourists and businesspeople were first allowed to visit Hong Kong and Macao under restricted conditions, and eventually to travel further afield.

Deng's policies of greater openness were also evident in education. Because one of the Four Modernizations was the impetus to develop Chinese science and technology, Deng recognized the need to access cutting-edge technology from around the world. He declared that China would send 3,000 students abroad every year.[200] These students returned to serve as the founding generation of Chinese academia, science, and engineering. More than 70 percent of Chinese university presidents and over 90 percent of faculty at Chinese national academies had studied overseas.[201] In 1980 China sent about 2,000 students overseas; this increased by the year 2000 to nearly 40,000. Before the COVID-19 pandemic hit, in 2019, China had over 700,000 students studying abroad.[202]

In sum, this chapter has explored the mechanisms of smart authoritarianism in China. Relative to China's highly extractive traditional society, Mao Zedong's more inclusive policies on women, marriage, and education created a demographic transition that laid the foundation for later growth. And during the reform years the CCP encouraged economic growth—by reducing repression, by opening the economy, by institutionalizing the government and constraining officials, by allowing civil society, and by raising the level of human capital. Smart authoritarian policies enabled China to defy the skeptics, fueling catch-up growth and then propelling China into the ranks of the world's most innovative countries.

Discussion

This chapter argues that smart authoritarianism fueled China's rapid economic growth after 1978 and (as shown in Chapter 2) has also enabled

China to become a global innovation leader. Critics might argue that the argument overemphasizes the role of Chinese leaders, when China's transformation was actually driven by its people.

Kate Zhou has argued against regime-centric views of China's economic rise to emphasize the agency and impact of Chinese farmers. "At the core of China's transformation," Zhou writes, "has been a rights acquisition movement undertaken by individual citizens who were fed up with the government controlling every aspect of their lives." This led to "the emergence of liberal practices that formed bases for future demands for freedom."[203] Zhou's argument is an important correction to a narrative that the Chinese people simply reacted to incentives provided by their leaders; she compellingly illustrates the dynamism and ideas of China's people, even while they were oppressed and starving. In this view, China's rise occurred because the CCP got out of the way.[204]

Just as an analysis that fails to acknowledge the creativity and striving of the Chinese people would be incomplete, so too would an analysis of Chinese growth that downplays the effects of the CCP's dramatic policy changes. First, as noted in Chapter 3, governments often do *not* get out of the way. It's thus vital to understand the significance of governments getting out of the way, to understand the mechanisms through which they do it, and to study the conditions under which they choose to do it.

Second, governments do much more than just get out of the way: They shift from old policies to new ones and orchestrate societal change in order to encourage economic development. This chapter, for example, showed that intentional government reforms caused China's demographic transition—an essential prerequisite to economic development. Chinese society had oppressed women for thousands of years and, absent Mao's revolutionary policies, showed every sign of continuing to do so. The government did not transform society alone; it partnered with activists (the All China Women's Federation) to achieve its goals.[205] But in changing the role of women in Chinese society, the government didn't get out of the way. It led.

Finally, government power is apparent in *reversals* of once-bestowed freedoms. China's people indeed pushed for more rights, and the smart authoritarian government relaxed many controls over a variety of areas, notably in civil society. But the Xi Jinping years show that while smart authoritarians give they also taketh away, affecting how free the Chinese people are allowed to be. In sum, the creativity, striving, and sacrifices of the Chinese people caused China's economic growth—as did the CCP's adoption of smart authoritarian policies.

We Should Excel in Technology

"Unless we change, the Westerners will be rich and we poor," wrote Chinese diplomat and reformer Xue Fucheng. "We should excel in technology and the manufacture of machinery; unless we change, they will be skillful and we clumsy. Steamships, trains, and the telegraph should be adopted; unless we change the Westerners will be quick and we slow . . . they will be strong and we shall be weak.[206] Xue was right: China's failed reform led to its decline from the great-power ranks, and to a period of weakness and turmoil.

A century later, Chinese reformers lamented that China had "slept" during the global economic transformation of the late eighteenth and nineteenth centuries, and in a "New Industrial Revolution" were determined not to repeat the error.[207] Observers correctly note that after 1990, Party leaders were preoccupied with the Soviet cautionary tale (i.e., the mistake of too much liberalization). But the Century of Humiliation, of China's nineteenth-century disasters vis-à-vis Japan and the West, also loomed large in Chinese leaders' minds.

This chapter has shown how the CCP's tempering of authoritarian viciousness enabled China to rise to become a great power and peer competitor of the United States. Mao is not known for governing China with inclusive policies. But traditional Chinese policies related to women and the family had been so staggeringly extractive that his reforms (e.g., maybe let's not cripple girls and marry them into polygamy at age 12) nonetheless had a profound economic impact, causing fertility to fall and instigating China's demographic transition. Subsequently, the CCP's smart authoritarian policies—for example, enacting institutional constraints, raising human capital, allowing civil society—fostered not only catch-up growth but even innovation at the global frontier.

While smart authoritarianism explains China's economic and technological rise, many observers question China's future because of a change in direction in CCP policies. Many scholars argue that Xi Jinping's "neo-authoritarian" turn will undermine China's future innovation and economic growth.[208] What is the future of China's smart authoritarianism—and of its economic and technological power? Chapter 6 examines this topic. Meanwhile, Chapter 5 turns to cases of smart authoritarianism beyond China.

CHAPTER 5

Crazy Rich Authoritarians

People are fascinated by dictators. In addition to a scholarly literature about authoritarian politics, books advise popular audiences on everything from "How to Be a Dictator" to "How to Feed a Dictator."[1] The "Dictator's Handbook" explained how leaders stay in power, while "How Democracies Die" warned against letting them accumulate power in the first place.[2] Critics of Donald Trump compared him to Adolf Hitler or Benito Mussolini in the lead-up to the 2016 and 2024 elections.[3] But one dictator to whom he was never compared was Singapore's Lee Kuan Yew. Indeed, people seem to forget that Lee was actually authoritarian.

Lee Kuan Yew ruled until his death in 2015 over a highly advanced economy governed by single-party authoritarian rule. Lee's success in doing what was thought impossible—staying in power as an authoritarian leader while fostering a rich, innovative economy—can be explained by his smart authoritarian policies. To be sure, as a tiny city-state of 5.7 million people, Singapore is a highly unusual country so perhaps can't serve as a model for others. However, China's success suggests that smart authoritarianism is more portable than many people expected.

This chapter looks beyond China to examine smart authoritarianism across history and around the world. I first explore how, in previous eras, dramatic economic, technological, and political change forced authoritarian leaders to adapt in order to compete with more liberal societies. I next turn to the case of Singapore, whose president, Lee Kuan Yew, invented modern smart authoritarianism: leading not only by his country's astoundingly successful example, but also through explicit guidance. From Singapore we follow Lee in his travels around the world—to Central Asia, the Middle East, and Africa—where he advised developmentally minded rulers about the smart authoritarian model. All of these cases suggest the appeal and broader viability of the model—which is bad news for the spread of democracy and human rights.

CHAPTER 5

Authoritarian Adaptation Over Time

Authoritarian leaders are famously chary of change. When an engineer presented Rome's Emperor Vespasian (69–79 AD) with a new and cheaper method of transporting heavy columns, the emperor rejected the idea. As quoted by the historian Suetonius, the Emperor protested, "How will it be possible for me to feed the populace? You must allow my poor hauliers to earn their bread." But at times, economic and military changes forced authoritarian leaders to adapt.

The First Industrial Revolution confronted authoritarian leaders with dramatic technological change that boosted economic output and increased military power yet threatened traditional tools of regime control.[4] With the Industrial Revolution and the factory model of production came a growing need for educated workers, increased mobility around the country, and increased urbanization. As noted in Chapter 4 with the example of the Qing, however, conservative autocrats worried about the social change that new technologies would create—and how this might affect political stability. Educating the public risked that the lower class "might be corrupted to aspire beyond its station," worried Prussia's Frederick Wilhelm III.[5] Austria's Emperor Francis I ruled out railroad construction, "lest the revolution might come into the country."[6] Austrian and Russian leaders also initially banned factory construction, fearing that factories would displace workers, creating unemployment and unrest—and worrying that activists would agitate and mobilize workers into antigovernment activity.[7] And—following Emperor Vespasian—China's Qing rulers rejected steamships because they would displace and potentially radicalize junk sailors—many of whom had previously risen up against the Qing in the Taiping Rebellion.[8]

Regimes that failed to absorb new technologies ended up significantly disadvantaged relative to their more adaptive rivals.[9] Productivity and economic output exploded in countries that laid more railroad track, adopted the factory model of production, and trained their people—by introducing state-run and universal education—to use and innovate with new technologies. Economic and technological strength translated to military advantage; countries that lagged technologically (e.g., Russia, Austria-Hungary, the Ottoman Empire, Qing China) experienced military disasters, which often incited political turmoil at home.[10]

Declining economic and military competitiveness led many authoritarian countries to overcome internal opposition and embrace reforms. Prussia's 1806 military catastrophe at Jena-Auerstädt—which lopped off half of Prussian territory and led the French to impose a massive reparations bill—fueled the Prussian Reform Movement. Fredrick William III and pro-reformist officials implemented political reforms, broke up guild monopolies, and abolished serfdom.[11] Education reformer Wilhelm von Humboldt—namesake of the famed university in Berlin—promoted the new view that not only elites but all people were "educable beings," and that the responsibility for public education lay with the state rather than the church.[12] By 1837, 80 percent of Prussian children (ages 6–14) were enrolled in primary education.[13] The Prussian state embraced railroad development, using it to tremendous effectiveness in the wars of German unification. Later, authoritarian Germany became a science and technology (S&T) powerhouse, building industries around chemicals, electricity, and the internal combustion engine.[14]

In Russia, economic crisis and military catastrophe in Crimea ushered in Alexander II's sweeping reforms.[15] Serfdom, specifically the policy of "ascription," had confined Russia's serfs to rural areas. The abolition of serfdom in 1861 enabled laborers to move to cities. Alexander II reformed education and sought to improve the flow of ideas. A national education system provided free and secular education in primary schools, and the lower classes could compete for spots at secondary schools.[16] Reforms also targeted Russian universities; Alexander loosened government control, granting them greater independence—such as control over their curricula (which under Nicholas I had been regulated by the government).[17] Alexander II also sought to improve the level of intellectual activity in Russia by lifting censorship, enabling Russian scholars to study foreign research. The government began issuing passports and encouraging foreign study, which previously had been prohibited.[18] Another important reform for Russian economic and military power was the country's embrace of the railroad. Alexander II oversaw massive railway expansion, implemented by reformer Sergei Witte; during this time Russian railroad track miles jumped from less than 700 to 14,000. This allowed a dramatic expansion in grain exports, which rose from less than 3 billion pounds to over 9 billion pounds between 1860 and 1880.[19]

Although wary of change, many authoritarian leaders discovered that reforms created new levers of control. For example, though elites had feared the prospect of educating the masses, leaders used education as a powerful nation-building tool and as a means of disseminating pro-regime ideas.[20] In Prussia, education reforms were conducted in ways to encourage conservatism and national unity. Officials "saw in schooling hope for controlling the social fluidity of early nineteenth-century Prussia"; conservatives viewed that more effective national education would inculcate a sense of national unity that would in fact reduce division and revolutionary tendencies.[21] In sum, before twentieth-century autocrats adapted to the challenge of the information age, some of their autocratic forebears had adapted to challenges of the industrial age.

The OG

The tiny city-state of Singapore was born in crisis. First a British colony, then occupied by Imperial Japan, Singapore in 1963 was incorporated into a federation with Malaysia. But only two years later it was expelled from this federation, becoming independent in 1965. Denied the common market it had been expecting and faced with the pullout of British military forces (which had previously funded 20 percent of GDP and provided many jobs), Singapore faced dire circumstances.[22] Per capita GDP was $300. The city-state had barely any land and no natural resources—even lacking fresh water to sustain its two million people. In 1960, notes Stephan Haggard, manufacturing comprised under 12 percent of the economy, and 94 percent of exports were re-exports.[23] Singaporean leaders remembered their conquest by the Japanese; they feared attacks from Malaysia and Indonesia. Their society comprised "a volatile mix of races and languages."[24] Thus leaders feared internal ethnic strife fanned by communist agitation. As the young politician Lee Kuan Yew commented, "We don't have the ingredients of a nation, the elementary factors." In his memoirs Lee recalled, "We faced tremendous odds and an improbable chance of survival."[25]

But under the rule of Lee and his People's Action Party (PAP), Singapore became a jaw-dropping economic success. It boasts one of the world's highest per capita GDPs in the world, which in 2023 was $84,734 in current $US—higher than the United States at $82,769.[26] In its first 25 years, Singapore's economic growth averaged 7.7 percent a year.[27] Singapore is a major exporter, its economy driven by value-added manufacturing (notably in the electronics and precision engineering sectors), in information and communications, and in finance and insurance. The World Bank ranked Singapore first in the world in terms of human capital; the Heritage Foundation ranked Singapore at #1 in the world in economic freedom. In its 2022 index, Heritage notes, "Singapore's highly developed free-market economy owes its success in large measure to its remarkably open and corruption-free business environment, prudent monetary and fiscal policies, and a transparent legal framework. Trade freedom is strong, and well-secured property rights promote entrepreneurship and innovation effectively."[28] Indeed, the World Economic Forum ranked Singapore first as the world's most competitive economy, and the country stands at second in the world on Bloomberg's Innovation Index.[29] Another global innovation index, the Global Innovation Index published by the World Intellectual Property Office, puts Singapore at #5 out of 132 countries (#1 in East Asia).[30] The country is an economic and technology powerhouse.

How did authoritarian Singapore do it? Prime Minister Lee Kuan Yew listened to Aristotle, tempering authoritarian viciousness.

DEMOGRAPHIC TRANSITION

Singapore's economic "miracle" began (as all growth miracles do) with a demographic transition. Following the drop in mortality taking place around the world after World War II, a subsequent drop in fertility expanded female labor force participation and promoted higher levels of human capital.[31] Partnering with civil society organizations, Lee's government engineered the demographic transition through inclusive policies that invested in public health, family planning initiatives, and universal public education.[32]

Women's rights were an important dimension of the government's effort to bring down fertility rates. The PAP had made a campaign promise to pass a Women's Charter, and it did so in 1961, banning polygamous civil marriages and legislating against violence in the family.[33] Reformers viewed equality and opportunity for women as a foundation for economic development. "Children, the spring source of the nation, had to be valued and no difference drawn between girls and boys. Both had to be nurtured and educated to the highest level possible. . . . In all this it became crucial that women be accorded full equal rights and protection as men."[34] More broadly, Lee Kuan Yew recounts of PAP reforms, "We shared the view of the communists that one reason for the backwardness of China and the rest of Asia, except Japan, was that women had not been emancipated. They had to be put on a par with the men, given the same education and enabled to make their full contribution to society." As a result of the ban on polygamy, increased educational opportunities, and the government encouraging later marriages, fertility dropped by half among girls and women ages 15–19.[35] The demographic transition enabled more women to join the formal workforce. From 1957 to 1992, female labor force participation rose from 21.6 percent to 51.3 percent.[36] A range of inclusive government policies thus touched off the demographic transition that created the foundation for Singapore's extraordinary economic growth.[37]

POLITICAL INSTITUTIONS AND CIVIL SERVICE

As in other competitive authoritarian countries, Singapore's government rules through ostensibly competitive institutions that in reality protect PAP dominance. A former British colony, Singapore was modeled on a Westminster-style parliamentary government. Elections for parliament began in 1959; though a small number of opposition parties do exist, the PAP has been in the majority since the country's founding. Notes Kenneth Paul Tan, "With an overwhelming majority in parliament, the PAP government has been able to amend the constitution without much obstruction, introducing multi-member constituencies, unelected parliamentary membership, and other institutional changes that have, in effect, strengthened the government's electoral dominance and control of parliament. With incumbency comes electoral advantages that have further secured the PAP's position."[38] Indeed, the PAP has changed the constitution and otherwise implemented policies to keep the opposition from gaining political power. Despite this, notes Cherian George, the PAP relies on "the legitimating power of elections, which are at least free and fair enough to attract the continued participation of all major opposition parties."[39]

One PAP innovation was the Group Representation Constituency (GRC), in which in some districts—chosen by the government—representatives to the legislature are elected not individually but by slate. The government explained this initiative as seeking to improve multiracial representation, but it advantaged the PAP by making it much harder for the opposition to gain seats. "It is hard enough for the opposition to attract candidates of any degree of calibre into its fold, let alone compile the right racial mix with its limited manpower resources."[40] The PAP also created a legislative position called the "nominated member of Parliament," a position with highly circumscribed voting rights. Ostensibly this sought to bring experts from various fields into parliament in order to benefit from their ideas and expertise. Through the initiative, the PAP sought "to steer disaffection with it away from the formal opposition in favor of co-optation": "These appointments [were] a preemptive move to ensure that any disaffection with the government from de facto interest groups [did] not translate into greater support for opposition parties."[41]

Singapore thus has a president, a prime minister, a legislature, and ostensibly free elections. But as Freedom House summarizes, "They are unfair due to the advantages enjoyed by the incumbent party, including a progovernment media sector, the GRC system, high financial barriers to electoral candidacy, and legal restrictions on free speech."[42] Through such policies and through low-intensity repression (described below), Lee and the PAP have governed since the country's founding. Lee's son, Lee Hsien Loong, became prime minister in 2004, and "made Singapore even more prosperous by largely following the semi-authoritarian and free-market model pioneered by his father."[43] He stepped down in 2024, handing power to his deputy, who continued PAP rule.

Professionalized Civil Service. Singapore's bureaucracy was created during colonization in the British model, in which bureaucrats were recruited by competitive examination and promoted on the basis of performance. Under Japanese occupation in World War II, however, Singapore's civil service featured high levels of bureaucratic corruption.

After independence Lee sought to professionalize the civil service. The government recruited top university students in order to attract "the best and the brightest." The PAP also tried to stamp out any corruption within the bureaucracy, such as the key legislation of the 1960 Prevention of Corruption Act (POCA). This significantly expanded the enforcement powers of the Corrupt Practices Investigation Bureau, imposing large fines and prison sentences on corrupt bureaucrats. As the country developed, the government raised civil service salaries, to keep pace with the private sector and to dissuade corruption; as a result of salary changes over the years, senior civil servants in Singapore earn perhaps the highest salaries in the world compared with their counterparts in other countries.[44] Singaporean government institutions—the Economic Planning Board, Development Bank of Singapore, and Monetary Authority of Singapore—comprised well-trained civil servants who supported Lee and the PAP in their effort to "formulate a set of coordinated government policies, mobilize savings and provide public goods."[45] This remains the case today. As Jeevan Vasagar notes, "There is a fierce struggle for the top positions in the country's government. Its ministers hold degrees from some of the world's finest universities, including Cambridge, Harvard, and Yale. Their qualifications range across hard science and social science."[46] Make no mistake, scholars note: Singapore's civil service remains closely supervised by the PAP, which approves all high-level appointments.

LOW-INTENSITY REPRESSION

Repression in Singapore dates back to British colonial-era anticommunist activities, and to Malaysia's Internal Security Act of 1963, which was enacted when Great Britain and Malaysia were joined and extended into Singapore after Singaporean independence. The act allows for indefinite preventive detention.[47] In Singapore's early years, the PAP relied on "secret police, detention without trial, and curbs on press freedom."[48] Decades later, although repression became less visible, "coercion, nevertheless, remains one of the pillars of PAP dominance," argues Cherian George, and the government retains the repressive Internal Security Act, which permits "arrest without warrant and detention without trial."[49] Over time, however, the PAP grew skillful at co-optation; furthermore, it gained performance legitimacy by overseeing economic growth, as did the government's narrative about the laudable—yet fragile—prosperity that the PAP provided.

The PAP pursues low-intensity repression to prevent opposition politicians from gaining power, as Chapter 3 described with the story of J. B. Jeyaretnam, or "JBJ." Opposition candidates receive disproportionately brief and negative media coverage, and the government often targets them with lawsuits for nonpolitical crimes. As noted, JBJ faced numerous lawsuits; he was found guilty and faced steep fines, which drove him to bankruptcy. When he was unable to pay further court fees, he had to serve a month in jail, leaving him disbarred as well as ineligible to stand for parliament. The tenacious Jeyaretnam successfully appealed this ruling to Singapore's Privy Council, which found him the victim of "grievous injustice"—to which the government responded by quietly eliminating a citizen's right to appeal to the Privy Council under such circumstances.[50] Another opposition politician who waged a similar battle was Chee Soon Juan, Secretary General of the Singapore Democratic Party. After Chee made critical statements about former prime ministers

Lee Kuan Yew and Goh Chok Tong during his 2001 campaign, the courts deemed these remarks libelous and commanded Chee to pay a fine of S$500,000.[51] Unable to pay, Chee declared bankruptcy and so, like Jeyaretnam, was barred from running for public office. Chee's sister and other fellow opposition leaders also faced steep fines.

The PAP also relies on low-intensity repression to control the media. Lee said in a 1971 speech, "Freedom of the press, freedom of the news media must be subordinated to the overriding needs of the integrity of Singapore and to the primary purpose of an elected government." Human rights organizations categorize Singapore as "not free"; in terms of press freedom it is ranked at 153rd in the world, in the same neighborhood as Afghanistan, Iraq, and Qatar.[52] But the government does not control information through force; the PAP "has achieved effective guidance of the press without either nationalizing ownership or brutalizing journalists."[53] Reporters, writes Jeevan Vasagar, are "pressured by their editors to ensure their political coverage never challenges the government."[54] As one journalist told Vasagar, "You're allowed to write what you want to write. After that, the editors will change it so much that you can't recognize it any more."[55]

The PAP also keeps the media in line through licensing and litigation. The Newspaper and Printing Presses Act requires newspaper publishers to have government permission in order to operate. At its inception, the *Straits Times*, Singapore's globally known newspaper, issued a formal editorial policy in which it stated its commitment to "getting behind policies that it deemed necessary for the development of the country."[56] Media companies must obtain a government license, which the government will withdraw if it disapproves of coverage. "While deprivation might be coercive, possession is a reward. Licencing increases the barriers to entry into an industry that is already hard to break into."[57] Furthermore, the PAP indirectly controls two of the most important media conglomerates, MediaCorp and Singapore Press Holdings Limited, and controls all broadcast TV channels.[58]

PAP media control enables the party, to use Steven Levitsky and Daniel Ziblatt's metaphor, to "tilt the political playing field." The PAP-controlled media provide less coverage to opposition candidates and cover them more negatively, which essentially relegates the opposition's campaigning to social media. But as social media has risen in importance, the PAP has cracked down in this realm too. Recently, the Protection of Online Falsehoods and Manipulation Bill (POFMA) is said to address "fake news" or misinformation; however, one journalist criticized it as "vaguely worded" and "wielded completely at whim by the government to suppress any form of dissent." He noted that "so far, POFMA directives have been little more than tools of pro-government spin."[59] The PAP's methods of control "include hiring social media consultants, 'Internet Brigade' and 'influencers' to soft-sell public policies and counter anti-establishment online comments."[60]

The PAP also relies heavily on lawsuits to control journalists and media organizations, filing crippling lawsuits that bankrupt those who step out of line. Under Singaporean law, libel is not merely a civil offense; it is a criminal offense punishable by up to two years in jail.[61] "The government appears to resort to criminal investigation or defamation lawsuits at every deemed affront," notes Human Rights Watch, "and it wields the laws in a manner so harsh that it creates a real sense of fear and has a chilling effect on critical speech." Media organizations have learned to self-censor in order to avoid attracting ruinous lawsuits. As one student activist told Human Rights Watch, the "genius of the system" is about creating "a culture of fear, rather than throwing people in jail."[62] At the same time, as Human Rights Watch argues, "beneath the slick surface of gleaming high-rises . . . it is a repressive place, where the government severely restricts what can be said, published, performed, read, or watched."[63]

CIVIL SOCIETY

Singapore's smart authoritarianism has permitted the growth of a large civil society, over which, as noted, the government maintains influence through low-intensity repression. A private media is well developed across diverse platforms, including traditional print and broadcast media; its vibrant digital media rely on a very high internet penetration rate of 94 percent.[64] News outlets such as the *Straits Times*, *South China Morning Post*, and *Channel News Asia* are regionally popular and respected. TV dramas produced by Mediacorp are beloved by Singaporeans and others in the region. Two public universities—the National University of Singapore and the Nanyang Technological University, Singapore—rank among the top 50 universities in the world and the top ten in Asia.[65] Furthermore, "a host of voluntary welfare organizations (VFOs) cater to specific cultural and social needs of the population. These nonprofit organizations provide basic social and welfare services to the elderly, youth, and the disabled." Civil society has encouraged debates and legislative changes on a host of issues; in addition to women's issues (described earlier), these included campaigns about the rights of migrant and domestic workers, living standards for the elderly, and issues related to the environment, conservation, and endangered species.[66]

As large and lively as Singapore's civil society is, however, it is circumscribed by government control. Suzaina Kadir argues that it's difficult to find where civil society and state begins: "Although these organizations are nonprofit and voluntary, the ties between such groups and the state remain close."[67] Government involvement, monitoring, and (as noted) low-intensity repression ensure that "civil society activism in Singapore does not involve confronting the state or undermining it."[68]

HUMAN CAPITAL

"Talent is a country's most precious asset," Lee Kuan Yew declared: "For a small resource-poor country like Singapore . . . it is the defining factor."[69] Thus the PAP invested heavily in education. Lee noted that under British rule, Singapore had benefited from "good schools, training for teachers, King Edward VII Medical College, and Raffles College" and that the country already attracted elites from Southeast Asia for training.[70] After independence, the first Five Year Plan (1961–1965) guaranteed free universal primary education, emphasizing math, science, and technical subjects.[71] The next several years saw rapid school construction, increased recruitment of teachers, and soaring enrollments in primary as well as secondary education.[72] Lee viewed education as imperative for raising the level of human capital, and additionally he saw schools as a vehicle for creating a sense of national identity to unify the new state's multiethnic population.[73] In the 1970s, concerned that the system was serving only the very top students and leaving others behind, the PAP sought to make education more inclusive.[74]

By improving educational access and quality, Singapore created a highly skilled workforce. "The educational standards of that work force were dramatically upgraded: While in 1966 more than half the workers had no formal education at all, by 1990 two-thirds had completed secondary education."[75] For decades, Singapore has topped the OECD's Programme for International Student Assessment (PISA) scores (which rank countries by math, science, and reading test scores), only recently to be overtaken by four Chinese regions (Beijing, Shanghai, Jiangsu, Zhejiang).[76] As noted, top Singaporean universities are highly respected both regionally and globally.

CAPITAL

Lee encouraged Singapore's economic development by aggressively promoting savings and investment. The government promoted capital formation via two former British institutions: the Central Provident Fund (CPF), which was government-mandated retirement savings, and the Post Office Savings Bank (POSB). The government reformed both in the 1960s, to great success: A decade after reforms, investment as a share of GDP had tripled to 33 percent.[77]

Singapore's government also cultivated foreign direct investment (FDI), which (as discussed further shortly) was a notable departure from a trend among developing and newly independent former colonies that viewed foreign investment as a tool of foreign domination.[78] Lee wrote in his memoir that his government "systematically built on Singapore's location and time zone advantages to promote the Republic as a regional and international financial center."[79] In their efforts to woo foreign capital, "Singaporean leaders sought large, stable multinational companies that would be committed to Singapore for a long period of time, and in order to attract such companies they gave very favorable conditions," such as building an industrial park.[80] Through a variety of policies, Singapore's government reassured foreign investors that Singapore was a stable, lucrative place to do business: "We established good labor relations and sound macroeconomic policies, the fundamentals that enable private enterprise to operate successfully."[81]

Singapore also attracted foreign capital because of the government's respect for property rights and the quality of its judiciary in the realm of commercial law. As noted, its courts are not known for their political independence; the PAP routinely uses the courts to weaken political opponents through libel, sedition, and defamation lawsuits. Yet the judiciary is formally independent, and efficient at litigating commercial disputes: convincing inventors and investors that Singapore is a stable society in which property rights will be upheld. Thus we confront what Jothie Rajah calls a "Singapore paradox: a regime that has systematically undercut 'rule of law' freedoms has managed to be acclaimed as a 'rule of law' state."[82] As Gordon Silverstein argues, "Singapore forces us to recognize the error so many Western politicians, pundits, and academics make in conflating liberal democracy—and its maximization of individual liberty—with the rule of law."[83]

OPENNESS

Fans of the hit book and film *Crazy Rich Asians* would probably be startled to hear that the story takes place in an authoritarian country. To most people, Singapore is "a glitzy international destination. It has hosted Taylor Swift concerts and Formula One night races."[84] "From Coldplay to Seinfeld to Ed Sheeran and even the Broadway musical *Hamilton*, it feels like every major act is coming to the Lion City."[85] One such act was the 2020 summit between Donald Trump and North Korean leader Kim Jong Un, which helped Singapore cultivate an image as a diplomatic hub.

In the early years after independence, Lee and his chief economic architect Goh Keng Swee (the country's first finance minister) positioned Singapore as an entrepôt nation, a "global city," in the words of its first foreign minister, Sinnathamby Rajaratnam.[86] As such the government not only cultivated FDI but also identified shipping and tourism as key sectors. The Port of Singapore Authority International processes "'about one-fifth of the world's total container transhipment throughput,' and was connected to 600 ports in 123 countries through a network of 200 shipping lines. This made the island republic the largest container transhipment hub in the world."[87] Singapore ranks in the top five of the world's most open economies to tourism.[88] This results in part from its fortuitous geographic location in rapidly industrializing Asia, one highly appealing to global investors and travelers. Such successes were far from inevitable, but rather resulted from Singapore building "competitive manufacturing and excellent infrastructure for attracting multinationals"[89] and working with other Southeast Asian leaders to shape tourist flows.[90]

That a poor former colony would pursue a model based on international openness seems unremarkable today, but it was a notable departure in the early 1960s. Swept up in decolonization, the developing world had embraced import-substitution industrialization (ISI), a growth model that, through protectionism, sought to build self-sufficiency and diminish foreign dependency.[91] Goh criticized the ISI model for the corruption it engendered, astutely arguing that it incentivized people to focus on "obtaining official permits and licences more than on efficient production and management."[92] Newly independent countries were also embracing socialist systems, and were defiantly severing relations with their former colonial masters. In 1958, the leader of Guinea, Ahmed Sékou Touré, sponsored a referendum to reject ties with France, saying "Guinea prefers poverty in freedom to riches in slavery."[93] In his 1964 tour of Africa, Lee met with Touré and the leaders of 16 other countries, and predicted poverty for those that were embracing socialism, rejecting free markets, and severing ties with their former colonizers and other influential capitalist economies.[94]

Goh and Lee's commitment to markets and openness infused every dimension of Singapore's nation-building, including the country's sense of national identity. For example, Lee chose to retain the prominent statue of Sir John Raffles—a symbol of British colonization—in order to avoid signaling revolutionary, anti-Western sentiment to foreign governments and investors.[95] This policy persists today. In an era in which capital cities have torn down graffiti-sprayed statues (e.g., "Rhodes Must Fall"), Singapore has actually added other Raffles statues, such as a marble copy to a vibrant area along the Singapore River, said to be Raffles's landing site. The national tourist office describes "quirky facts" about the statue, refers to him by his nickname of "Iron Man," and encourages visitors to "take your selfies with this popular figure."[96]

Openness was seen as essential for creating a highly productive workforce and innovative economy. Notes Elaine Ho, "The globalization of talent is of particular concern for the Singapore government—both inflows of foreigners to Singapore and outflow of Singaporeans to other countries—because human capital is treated as a key determinant of the country's competitiveness."[97] Toward this end, Lee decreed that the "global city's" language of public education would be English, which both helped unify the multilingual society and enhanced Singapore's ability to join global education and knowledge networks.[98]

In sum, Singapore has followed (really, written) the smart authoritarian's playbook for the modern era, creating a system in which an authoritarian government managed to stay in power while overseeing the creation of a prosperous, innovative country. Singapore's most important export was the smart authoritarian model itself.

A World of Smart Authoritarians

Lee Kuan Yew actively promoted the spread of smart authoritarianism by traveling the globe to advise authoritarian rulers. Lee took 33 trips to China, where he advised aspiring smart authoritarian Deng Xiaoping, and also welcomed Deng to Singapore in 1978. Robert Kaplan argues, "For decades Western liberals consoled themselves by assuming that Singapore was an odd exception, a city-state of dynamic overseas Chinese, lacking a hinterland which made governing easy."[99] But all over the world, autocracies and backsliding democracies are following the smart authoritarian model to cultivate economic growth. Although a large body of literature on domestic institutions dismisses these countries' chances of ever becoming (or in Hungary's case, remaining) innovative economies, China's success suggests otherwise.

CENTRAL ASIA

In Kazakhstan, former President Nursultan Nazarbayev enthused about Lee's "extraordinary personality, forceful delivery, and great charisma." He described how Lee advised him during his 1991 visit about Singapore's path to economic development and "forecast a great future for Kazakhstan."[100] In 1995, the newly independent country adopted a new constitution and implemented significant political reforms. Kazakhstan also undertook major economic reforms aimed at privatization, price liberalization, and the creation of an attractive investment climate for FDI.

Education reforms were a major part of Kazakhstan's modernization push. Because the Soviet central government had previously provided significant education funding, the transition to independence gutted Kazakh education. The Soviet education system there had been underfinanced, highly ideological, and isolated from global networks, making the country ill-prepared to participate in the world economy or foster innovation.[101] In 2001, the government implemented its Education Reform Strategy, which "specified assurance of resources, improved quality of outcomes and the efficiency of the education system as key objectives, stressing that the quality of education . . . should be comparable to that of the developed countries of the world."[102] Higher education in particular was increasingly privatized and restructured to be in line with international structures (e.g., a four-year instead of five-year university program). With such reforms the Kazakh government aimed to increase "the competitiveness of education and [the] development of human capital" in order to promote sustainable economic growth.[103]

Reforms boosted Kazakhstan's economy. The World Bank notes that after a recession during its transition in the 1990s, the economy grew due to oil, gas, and mineral exports as well as market-oriented reforms and strong FDI.[104] Economic growth topped 10 percent between 2000 and 2006; in 2023 it was over 5 percent.[105] The country's GDP per capita rose from $3,700 in 1995 to $11,300 in 2022, lifting the country from the low- to upper-middle-income category.[106] Leaders announced plans of "Digital Kazakhstan" and "Kazakhstan 2050": plans that position innovation and digitalization as central to the country's long-term economic growth.[107] These programs, and efforts to close the rural-urban connectivity gap, are partly funded by the World Bank.

Following in the PAP's footsteps, Nazarbayev and his Nur Otan party relied on low-intensity repression to stay in power. Elections were held, and reforms enthused about "new, fairer, and more competitive rules for forming the representative branch of government."[108] But opposition candidates tell a different story—about onerous registration requirements and unequal media coverage. "We nonestablishment candidates," commented Lukpan Akhmedyarov, "are constantly looking at each other and asking each other when will we be ejected from the race and on what basis."[109] In 2022, President Kassym-Jomart Tokayev announced reforms aimed at creating a "New Kazakhstan" that included constraints on the executive and electoral reforms. These "helped him craft a reputation as a reformer in the West," while critics argue very little has changed—and Tokayev remains firmly in control.[110]

Kazakhstan's civil society has grown significantly over time, but the government retains significant influence over its activities. The government has responded to protests by fashioning itself as a "listening state," setting up a National Council of Public Trust to dialogue with civil society.[111] Yet participation is highly contingent on staying in the government's good graces. Civil Society Organizations (CSOs) must register with the government or face fines, closure, or even criminal prosecution.[112] Registration is a laborious process, and organizations often find themselves denied because of typos in their applications.[113] The Kazakh government also funnels funding from natural resource profits toward compliant CSOs, creating what Colleen Wood calls a "manicured civil society."[114] The government also oversees the media. Compared to 1991—when Kazakhstan had only 600 newspapers and magazines, all state-run—the country's media landscape has flourished.[115] But the government argues that it must monitor the media in order "to prevent damaging effects on society's moral development, as well as disruption of the universally humane, national, cultural, and family values."[116]

Another Central Asian country, Uzbekistan, has similarly followed a smart authoritarian model, in which "authoritarian repression does not disappear, but is usually softer and more subtle."[117] As one observer comments, Uzbek leader "Shavkat Mirziyoyev is trying to modernise, rather than liberalise, Uzbek authoritarianism. His agenda of authoritarian modernisation is oriented towards the achievement of rapid economic growth," through the abandonment of autarky in favor of globalization, and through regional and global connectivity.[118] Recently, Uzbekistan has experienced growth rates of 8 percent "by relentlessly pursuing the path of market-oriented reforms."[119] Smart authoritarianism is thus thriving in Central Asia.

MIDDLE EAST

Smart authoritarianism is also transforming the Middle East. "There is a "growing awareness in the Arab world as there is elsewhere that developing viable, competitive economies requires a host of reforms."[120] Leaders in the Arab Gulf States saw oil prices fall after 2014 and foresaw a global economic shift away from hydrocarbons. The Arab Spring protests of 2010–2011 also "forced the Saudi government to confront the fact that rising youth unemployment could no longer only be addressed by providing public-sector jobs."[121] To promote economic development and innovation, petrostates have announced strategic plans (e.g., "Qatar National Vision 2030," "UAE 2050")[122] to foster innovation and lift per capita GDP to high income levels.[123]

"Go to Riyadh," urges John Hannah, marveling at the "Vision 2030" reforms of the young Crown Prince Mohammed bin Salman (MbS). The transformation in Saudi Arabia, Hannah argues, "is one of the most important but underappreciated developments of the past decade, with profound implications for the Middle East and beyond."[124]

Saudi Arabia remains an absolute monarchy, with one of the world's lowest scores for political freedom. "There is no tolerance for any dissent, including on social media where users are watched closely using surveillance technology."[125] The regime cannily responded to the threatening rise of social media by "transforming Saudi Twitter into a platform populated by pro-regime influencers and automated 'bots' creating the illusion of popular regime support."[126] The cult of MbS, writes Madawi Al-Rasheed, "became sacrosanct, as any critical voices were either silenced by house arrest or put in jail.[127] Critics who fled the country became targets of intimidation campaigns abroad." Executions—including mass executions—have increased.[128] The world was shocked by the Saudi government's abduction and grisly murder of journalist Jamal Khashoggi in Turkey. This was a discordant display of authoritarian viciousness that contrasts with—and has severely set back—Riyadh's efforts to reshape its image.[129] Lee Kuan Yew would definitely not have approved.

Amid continued repression, however, Saudi Arabia's overall freedom score is actually rising because of improvements in economic freedom.

Comments Arezki, "The notion that economic transformation can happen independently of political transformation is certainly taking a page out of China's book."[130] Indeed, various reforms seek to attract FDI: "As a result of a new set of laws to promote entrepreneurship, protect investors' rights, and reduce the costs of doing business, new investment deals and licenses grew by 95 percent and 267 percent in 2022, respectively."[131] Furthermore, to prepare the workforce for innovative-based growth, the government is devoting significant resources to education reforms, which aim to increase years of educational attainment and to improve the quality of teaching and curricula.[132]

Saudi Arabia is also opening up to the world in numerous ways. Whereas prior to 2020 tourists were unable to visit, the government created an efficient system for administering visas; foreign tourism soared to 100 million tourists (seven years ahead of schedule). As the UN Tourism Office praises, Saudi Arabia experienced "a staggering 390% increase in demand for tourism activity licenses in 2023" and tourism contributes more than 7 percent of the non-oil economy.[133] Saudi Arabia joins other Gulf countries (and Singapore) in crafting a cosmopolitan identity through "sportswashing"; as the *Economist* notes, Saudi businessmen have purchased prestigious sports teams, brokered a "dubious new golf tour," and offered "a colossal contract for Cristiano Ronaldo, a Portuguese footballer."[134]

Another important dimension of reform is the shift—as from Mao to Deng—from ideological extremism toward economic pragmatism. MbS has "openly set about sidelining the kingdom's Wahhabi scholars and preachers who still command millions of followers in the country and beyond."[135] This sect advocates autarky, hostility toward the West, discrimination against religious minorities, and strict control of women (a familiar recipe for economic underdevelopment). MbS is taking the country in the opposite direction: allowing women greater freedoms and autonomy, such as allowing them to drive, allowing coeducation, and permitting young men and women to socialize together at theaters, dance clubs, and concerts.[136] In 2023 a female Saudi engineer joined a mission to the International Space Station.[137] But change extends beyond such high-profile events; Saudi Arabia's female labor force participation rate more than doubled since 2017, from 17 percent in 2017 to 36 percent in 2023.[138]

So far, Saudi reforms have generated significant growth in the non-oil sectors, at about 5 percent in 2022. Silicon Valley firms—OpenAI, Microsoft, and venture capital firms—are flocking to the Kingdom of Saudi Arabia, which is seeking to fashion itself into an AI leader. "The Khashoggi era," commented one venture capitalist, "is over."[139] Partnering with the firm Andreessen Horowitz, the Saudi government is creating a $40 billion technology fund that would make the Kingdom the world's top AI investor"[140] As he pursues reforms and seeks to foster innovation, as a Saudi government official commented to an interviewer, "MBS has Lee Kuan Yew imprinted on both sides of his brain."[141]

Discussing Vision 2030, Saudi Finance Minister Mohammed al-Jadaan commented at Davos in 2023, "We wanted to be a role model for the region and we are encouraging a lot of countries around us to really do reforms."[142] Indeed, other countries in the region are embracing the smart authoritarian model. In the United Arab Emirates (UAE), reformist leader Mohamed bin Zayed (MbZ) similarly seeks to encourage FDI and innovation to improve education and professionalize the civil service. Like Saudi Arabia, the UAE is reining in the political influence of extremist Islamists, opening its economy to Western investment and tourism, and making a major AI push.[143] And like MbS, the Emirati leader relies on extractive political institutions to do so, creating, as a *New York Times* profile noted, "a socially liberal autocracy, much as Lee Kuan Yew did in Singapore in the 1960s and '70s."[144]

In Jordan, reliance on smart authoritarian tools has enabled the monarchy to stay in power while presiding over economic growth. Before 1989, the regime ruled in blunt authoritarian fashion: dominated by the monarchy, through heavy repression and with weak institutions.[145] The economy was in shambles, with high unemployment and severe indebtedness that in 1989 created a national crisis. King Abdullah announced democratic reforms and a new era of pluralism. But democracy was not the plan; King Abdullah II (who took over after his father's death in 1999) ruled in the smart authoritarian model. He "staffed his government with technocrats whose mission was to push neoliberal reforms far enough to spark economic growth, but not so far as to put political stability at risk."[146] These technocrats presided over Jordan's global economic integration; Jordan built Qualifying Industrial Zones and Special Economic Zones and signed free-trade pacts with the United States and the European Union. "The state poured money into tourism, . . . speeded the privatization of state-owned enterprises, and aggressively sought foreign investment." All of this produced economic growth at an average of 7 percent from 2001 to 2008.[147]

Abdullah II has managed his country's economic liberalization while relying on low-intensity repression. The regime rode out the regional turmoil of the 2010–2011 Arab Spring with greater stability relative to its neighbors.[148] Civil society is kept in check through the familiar methods of registration and permits, monitoring, and nonpolitical lawsuits against activists whose behavior is viewed as crossing a line.[149] Stefanie Eileen Nanes, who examined civil society activism opposing violence known as honor crimes, argues that Jordan's government cleverly co-opts civil society activism.[150] "Subtle co-optation, rather than coercion, was the regime's most effective tool against even a small assertion of independent associational activity that tackled a sensitive social issue." Sean L. Yom agrees:

Soft mechanisms of manipulation have enabled Abdullah's regime to bolster its base and fend off large-scale opposition without having to resort to widespread violence. The manipulation strategy, moreover, allows the monarchy to maintain a veneer of political openness and moderation that allows Jordan to pose (with a special eye on Western donor countries) as a modern and relatively progressive polity amid the surrounding turmoil of the troubled Middle East.[151]

Overall, several authoritarian governments in the Middle East are recognizing a need to move away from an oil-dependent economy and are modeling their policies on the Chinese and Singaporean cases. As Calvert W. Jones argues, they are "calling for their citizens to become creative and critical thinkers, educated to the highest international standards, and trained to be innovators, knowledge workers, and social entrepreneurs."[152] Reforms have significantly helped their external relations: "The young monarchs of Bahrain, Jordan, Morocco, and Qatar . . . became darlings of the U.S. administration after 2001 in part because of their reformist credentials and benefited from free trade agreements, assistance packages, and enhanced military cooperation."[153]

AFRICA

Morocco's King Mohammed VI "initially set the country's course on a model of both political opening and socioeconomic development."[154] His 2021 New Development Model outlined socioeconomic goals and development targets for the country in 2035. The government built infrastructure, privatized the public sector, and expanded manufacturing; GDP per capita rose from $1,950 to $3,300 in 2022.[155] "All of this has contributed to shaping a vision of the country as a leading regional economic power."[156]

The Moroccan government also engaged in political reform—navigating protests associated with the 2011 Arab Spring as well as the Rif and Jerrada movements of 2016–2017.[157] Mohammed VI's father King Haman II had presided over the brutal "years of lead" in which political opponents were terrorized by disappearances, police raids, torture, and imprisonment in secret prisons.[158] Mohammed VI relaxed censorship, increased the fairness of elections, advanced the legal equality of women, and even acknowledged the human rights violations that had taken place during his father's reign. Yet "power in Morocco is still where it has always been, firmly in the hands of the king or, more broadly, the palace."[159]

Beyond Morocco, other African governments are presiding over impressive economic growth and rising living standards while tilting the political playing fields to stay in power.[160] In Uganda, President Yoweri Museveni has ruled for more than three decades, relying on smart authoritarian tools to extend his rule while presenting an attractive image to the global liberal community. Uganda has free elections—except that opposition candidates encounter significant obstacles, including a lack of access to media coverage. As Lynette Ong explored in China, Museveni's government outsourced the use of violence to vigilante groups (that it also supported) to distance itself from violence.[161] Toward civil society, the government pursues a strategy of "quietly cutting off and choking any organization that might challenge the government's authority"; troublemakers are slapped with penalties for crimes such as defamation, tax fraud, and money-laundering.[162] These smart authoritarian strategies enable the regime to stay in power while being "lavished with development and security assistance": receiving almost a billion dollars a year from the United States, "while another billion flows from other donors and global institutions such as the World Bank."[163]

In Ethiopia, observers once described leader Abiy Ahmed as Africa's MbZ or Deng Xiaoping.[164] Abiy took power in 2018 on a wave of reformist hope—privatization, an alphabet soup of development plans, a peace agreement with Eritrea—and won accolades from the international community. FDI poured in and Abiy accepted the 2019 Nobel Peace Prize.

Ethiopia's economy grew phenomenally: "one of the strongest growth rates in the world," averaging over 9 percent from 1999 to 2019. The government privatized former state-owned enterprises in aviation, telecommunications, and energy. The agricultural sector's share in the economy has decreased steadily, with services and manufacturing on the rise. "In industrial parks scattered across the country factories sprang up, many dedicated to making the textiles and clothing that often represent the first rung on the industrialisation ladder."[165] And under the African Growth Opportunity Act (AGOA), Ethiopia enjoyed duty-free access to the US market. Poverty rates have fallen and performance on human development indicators is rising.[166]

Smart authoritarianism in Ethiopia dissipated amid continued problems with Eritrea, internal ethnic violence, and a bloody civil war against Tigrayan rebels. The government responded to separatist activity in the Oromo region with arrests, torture, summary executions, and indiscriminate repression.[167] Furthermore, in the war against Tigrayans, NGOs reported gross human rights violations by Ethiopian forces, mass graves, and suspicious activity by the government to conceal bodies.[168] Abiy's government tightened political control, targeting the media and the political opposition.[169] The surveillance state has grown significantly—reliant on Chinese surveillance technology and on Chinese training of government officials.[170] For all these reasons, "Abiymania" cooled, with some costly repercussions; for example, the United States disqualified Ethiopia from the AGOA program because of human rights violations.

Yet Ethiopia's GDP continues to rise, a pivotal demographic transition is occurring, and investment keeps pouring in—in fact, attracting the region's highest amount of FDI. Ethiopia's capital, Addis Ababa, is in the midst of a massive demolition and construction campaign; "Abiy dreams of a city of tourism and technology, of grand parks and gleaming museums, overlooked by a planned multibillion-dollar complex which will include a national palace and a grand hotel."[171] His journey appears rougher than Lee Kuan Yew's, but Abiy is drawing from a similar toolkit—and hoping that the world community will see his country as "Africa's best hope of replicating the export-led growth of Asian states like Vietnam or Bangladesh."[172] Importantly, unlike many of the countries this chapter has described, Ethiopia has a population of over 100 million people. Its successful application of smart authoritarianism—though far from assured—could propel it to great power.

Lee's Legacy

In the modern, information-age, global economy, innovation-based growth under authoritarian rule was previously thought impossible. Singapore—a country that managed to foster innovation under authoritarian rule—was thought to be a tiny exception that proved the rule, not a model for other countries to emulate. But Deng Xiaoping, Nursultan Nazarbayev, Sheikh Zayed, and other autocrats disagreed. All of them invited Lee Kuan Yew to visit their countries, and all of them have followed Singapore's model. Their regimes are presiding over economic growth while maintaining power with the use of low-intensity repression, increasingly relying on digital tools of control. Whether these countries will succeed in emulating Singapore's and China's success in innovation remains to be seen.

Smart authoritarianism also appears in countries experiencing democratic reversals. Whereas the cases noted above are formerly strict authoritarian governments loosening political control in order to foster growth, the opposite also occurs: namely, formerly democratic governments (e.g., in Georgia and Hungary) tightening political control while still seeking to foster an innovative, globally integrated economy.[173] All over the world, then, we see a growing number of governments that seek to maintain (or impose) political control while tactically freeing their economies.

CHAPTER 6

Staying Smart

The man at the podium wore a dark suit and a red tie. Behind him sat rows of dignitaries in front of a vast wall, draped in gold, from which protruded a yellow hammer and sickle, framed on either side by 100-foot scarlet flags. In front of him, in the cavernous, red-carpeted hall, sat more than 2,000 delegates to the 20th People's Party Congress, at row after row of wooden desks. They listened attentively and took notes like their lives depended on it, which they may well have. The man, Xi Jinping, spoke for over two hours, during which his carefully rapt audience occasionally erupted in ecstatic applause. The moment was both bland in its authoritarian predictability—and a sea change in Chinese politics.

This was the occasion at which Xi declared himself, in defiance of term limits, China's leader for life. Over the years, Chinese Communist Party (CCP) leaders had turned the country in the direction of smart authoritarianism in order to lift China out of the poverty and chaos caused by Mao Zedong's more extractive rule. But many experts argued that Xi—flouting term limits, centralizing power, and doubling down on statist economic policies—was leading China back in the wrong direction.[1]

This book has argued that China has risen to great power and has become one of the world's most innovative countries. Its technological rise has surprised many observers, who have argued that, in an information age, extractive institutions will stifle innovation in authoritarian regimes. A prominent literature expects that authoritarian leaders face an inexorable tension between the extractive institutions that keep them in office and the inclusive institutions necessary to foster innovation. However, this view overlooks authoritarian heterogeneity and adaptability. China's economic and technologic success stems from its pursuit of smart authoritarianism.

Smart authoritarians, in response to late-twentieth-century trends that advantaged liberal societies, recognized the tension between political control and the conditions needed to foster growth and innovation. Following Aristotle's advice to become "half-vicious," they adopted more inclusive political and economic policies. As I show in Chapters 2 and 4, after the death of Mao Zedong, China's shift to smart authoritarianism not only fueled 40 years of stunning catch-up growth but also facilitated the country's rise to being a global technological leader.

What are China's future prospects? Several economists are cautiously optimistic, arguing that, contingent on several reforms, China will still experience significant growth. "China is still rising," contends Nicholas Lardy.[2] C. Fred Bergsten argues that "China is highly likely to acquire the capability to exercise global economic leadership across a wide range of activities over much, if not all, of the world," and that even a "modest growth profile . . . would bring it comfortably into a global leadership position on virtually all of the GDP, wealth, trade, technology and other metrics."[3] The most bullish observer was former Singaporean president Lee Kuan Yew. "The size of China's displacement of the world balance is such that the world must find a new balance," Lee commented in 2013. "It is not possible to pretend that this is just another big player. *This is the biggest player in the history of the world.*"[4] Many other observers, however, express deep skepticism about China's future economic growth.

This chapter probes the debate about China's future, and offers two key arguments. First, many people argue that China's growth will slow, and they identify a variety of reasons why—commonly referred to as "headwinds" to future growth. But economists argue that catch-up growth naturally slows at the middle-income stage, so of course China's growth will slow (for precisely those reasons). China's slowing growth is neither surprising nor grounds to expect its decline as a great power.

Second, I argue that the future balance of power depends on the CCP's ability to manage the tensions of the smart authoritarian model. The key question about China's future is neither whether China will maintain high growth nor whether it will evade the "middle-income trap." Rather, China's ability to remain a superpower depends on its continued ability to compete at the technological frontier—which depends on the CCP's continued management of the tensions between authoritarian control and the freedoms and openness of a modern, globalized economy.

Brush Clearing

China's ability to maintain its geopolitical position, and to compete effectively against the United States, depends on the CCP's ability to sustain economic growth (albeit at a much-reduced level), to continue to operate at the technological frontier, and to continue to mobilize military power. Pessimists make a variety of arguments about China's future, but some of these arguments are not relevant to the future balance of power.[5]

SLOWING GROWTH

First, China's great-power future does not depend on continued high growth. The slowing of China's economy (already occurring) is predictable. Economists explain that after a rising economy reaches approximately middle-income level, GDP growth settles down to about 1 to 2 percent per year.[6] As Chapter 1 described, this would reflect a successful shift from input-led growth (which fuels a country's rise to middle income) to innovation-based growth (which fuels growth past middle income).[7] In short, when debating China's future, the question is **not** whether China can overcome its numerous challenges to sustain 7 or 8 percent growth. No other middle-income country has ever achieved this and China won't either. Rather, the question is whether Chinese growth will successfully settle into sustainable levels at the 1–2 percent range.

CHINA AND THE MIDDLE-INCOME TRAP

China's future as a great power does not depend on it escaping the "middle-income trap."[8] Countries stalled in this trap fail to shift from catch-up to innovation-based growth. As a result, they are "unable to compete with low-income, low-wage economies in manufactured exports and unable to compete with advanced economies in high-skill innovations."[9] Whether China can escape the middle-income trap is a debate about whether its per capita GDP will continue to grow into the high-income category. For 2023, the World Bank put this threshold at $13,846 in nominal terms. At this writing, China (at about $13,400) is an upper-middle-income economy.[10]

A country's income status may be a relevant factor in discussions about economic development and standards of living, but it's less relevant in geopolitical debates about the balance of power. To begin with, the arbitrary World Bank threshold for "high income" of about $14,000 overlaps poorly with the world's most advanced economies. The United States, for example, logs in at $80,000, Germany at $51,000, Japan at $39,000.[11] There are far more high-income economies than there are great economic powers. More importantly, as I've argued elsewhere, GDP per capita statistics are poor indicators of great power. Some great powers have had low GDPs per capita— and plenty of non–great powers have had high GDPs per capita.[12] What drives the balance of power is whether a country has the demographic, economic, and military **scale** (which China has) and the technological **sophistication** (which China has) to compete. GDP per capita tells us little about either. The Soviet Union never made it past middle-income level yet was regarded as a superpower; it engaged the United States in a four-decade-long security competition that people feared would end life on earth. Today's upper-middle-income China is already more powerful than was the Soviet Union on most dimensions.[13] Although debates about China's per capita GDP are relevant to the quality of life of the Chinese people, they are less relevant to debates about the country's future as a great power.

TAILWINDS TO HEADWINDS

Commentators often highlight various tailwinds that boosted Chinese economic growth, which over time have shifted to headwinds, making future growth more challenging.[14] As I discuss shortly, these are indeed the mechanisms through which rapid economic growth slows. However, slowing growth is distinct from great-power decline.

A first headwind relates to China's increasingly unfavorable **demographics**. As Chapters 4 and 5 discussed in the cases of China and Singapore, the precursor to a country's economic rise is a demographic transition: a drop in mortality and fertility.[15] Such transitions produce a "bulge" generation (larger than the cohorts before and after it).[16] When this generation ages into the workforce, with the right economic policies it can create a demographic "dividend" that stems from having a larger number of workers and savers relative to dependents. However, as the bulge generation ages out of the workforce, the dividend shifts to a penalty, as the smaller cohort of workers must support the aging large cohort.[17] The CCP's One Child Policy exacerbated the dividend by severely suppressing the number of dependents; likewise, as the bulge generation ages out of the workforce to be replaced by an extra-small-size cohort, this will exacerbate the penalty.[18]

A second headwind relates to **environmental damage**; observers have argued that China must deal with environmental degradation caused by its rapid industrial rise.[19] With the emergence of a more prosperous middle class, people will no longer tolerate pollution of the water, soil, air, and other elements, or the disease and death that the pollution causes. More affluent people are more likely to demand environmental cleanup, and a more affluent country is better equipped to supply it.[20] Some observers have argued that in China, environmental cleanup will reduce economic growth.[21]

A third headwind is the risk of **financial crises**. In the property sector, which represents a quarter of China's GDP, "a credit bubble of historic proportions that drove China's growth over the past decade is currently unwinding, and slowing the economy as a result."[22] Observers have long warned of a debt crisis, given highly leveraged state-sector firms.[23] In addition to these domestic woes, China is facing an overseas debt crisis as countries in the Global South default on loans extended as part of China's Belt and Road Initiative.[24] Such borrowers were hard hit by the COVID-19 epidemic and rising food prices in the wake of Russia's invasion of Ukraine.

All of these headwinds are well-established causes of slowing growth as a country approaches middle income. Indeed, all these factors contributed to the growth slowdowns of previous rising powers. All rising economies—

previously boosted by the tailwind of a demographic dividend—later paid a demographic penalty. China's demographic challenges are significant and will indeed dampen growth.[25] But leaders also have tools they can use to address this issue, such as lengthening China's young retirement ages (60 for men and 50–55 for women in urban areas).[26] Furthermore, all rising economies wrecked their environments on the way up and—in response to a more prosperous and demanding middle class—later engaged in regulatory reform and devoted resources to environmental cleanup.[27] Many rising economies experienced financial crises, including Japan, South Korea, and the United States (which saw multiple financial crises in the nineteenth century during its economic rise). Yet today those countries rank among the world's richest and most technologically advanced.

In sum, pessimistic views of China's economic future are often based on arguments about the headwinds it faces. These headwinds are indeed why growth slows in what had been rapidly rising economies; they are indeed challenges that Chinese leaders must address. As Lardy reminds us, "China overcame even greater challenges when it started on the path of economic reform in the late 1970s."[28] And importantly, slowing growth does not necessarily translate to China's decline.

Solving the King's Dilemma

The CCP indeed faces many challenges going forward. But to compete against the United States in a great-power competition, China needs neither to reach the high-income category nor to surpass the United States economically or technologically. Rather, China needs to remain at the global technological frontier. This will require the CCP to continue managing the tensions of the smart authoritarian model—a feat that many skeptics doubt the regime can perform.

According to the model, the CCP must continue managing the tensions between political control and economic growth. At China's current developmental stage, this means the CCP must stay in power while providing the conditions that foster innovation. But holding onto power while, for example, creating a highly educated populace, allowing a dynamic civil society, and actively engaging with the world is a challenging feat for an authoritarian regime. Indeed, the feat was deemed impossible by institutions theory, modernization theory, and the architects of the US engagement policy toward China.[29] But smart authoritarianism argues that autocrats can stay in power and foster innovation by calibrating levels of political control and levels of political openness and freedoms: sometimes tighter, sometimes looser in response to constantly changing conditions.

This kind of calibration requires high-quality information about the economy, about public opinion, and about the level and nature of political opposition. It requires leaders (or their advisors) to understand the sources of innovation and to keep up with new technological developments and new methods of control. Not only must dictators access information; they must also analyze it competently or must surround themselves with advisors who can.

Skeptics might question whether smart authoritarian regimes can possibly be—or remain—this smart. Scholars have argued that dictators lack high-quality information.[30] In democracies, a variety of institutions (e.g., civil society, the judiciary, political parties, and the opposition) provide information about societal problems, the extent and nature of political challenges, and the state of the economy. Although even democracies confront informational problems, many scholars expect better information flows within democracies, and more and better information available to leaders.[31] Furthermore, observers argue that over time dictators become more narcissistic, paranoid, and/or isolated, which reduces their access to information, or affects their judgment of it.[32] In this view, smart authoritarianism should be rare—and if it occurs at all, it won't last.

Contrary to this view, a large literature on authoritarian politics has shown that many authoritarian regimes, recognizing the need for information, emulate democracies in order to obtain it. As discussed in Chapter 3, autocrats created judiciaries, legislatures,[33] and civil society with private media, universities, and other organizations.[34] China skeptics have neglected to consider how, following the smart authoritarian model, China and other autocracies have adapted in order to mitigate their informational disadvantages.

Not So Smart?

China in the early 2000s is remembered as being flush with energy and exploring new freedoms and openness. "Individual Chinese," observed George Gilboy and Eric Heginbotham, are able to travel abroad; they "are now free to create their own lifestyles: they can move about the country, start their own businesses, and express themselves on a wide range of issues."[35] Later, however, beginning under Hu Jintao and then deepening under Xi Jinping, the CCP imposed greater state control over the economy, civil society, and personal freedoms. The vibrance of the previous era has stilled, writes Evan Osnos; "The space for pop culture, high culture, and spontaneous interaction has narrowed to a pinhole. Chinese social media, which once was a chaotic hive, has been tamed, as powerful voices are silenced and discussions closed."[36]

Many observers argue that Xi Jinping is tightening up because of the Soviet cautionary tale and the dangers of *perestroika*: the risks that allowing too much freedom will ultimately topple the regime.[37] Such fears are leading Xi to "overreach," as Susan Shirk argues, in the opposite direction, toward stultifying control. Many observers warn that Xi's neo-authoritarianism will undermine Chinese growth and innovation—leading China down the Soviet path. Geremie Barmé comments, "The irony is that he, by his actions to make himself supreme leader, is in fact repeating the cycle of history."[38] As Osnos writes, "To spend time in China at the end of Xi's first decade is to witness a nation slipping from motion to stagnation and, for the first time in a generation, questioning whether a Communist superpower can escape the contradictions that doomed the Soviet Union." As I discuss below, observers point to policies such as a state-directed innovation model, the abandonment of state-sector reform, a return to personalist rule, and increased repression, all of which observers argue will reduce innovation and thus undermine China's future economic growth.

STATE-DIRECTED INNOVATION

Commentators argue that Xi Jinping is muscling the Party into a position of greater control over China's state, economy, and society. As Margaret Pearson, Meg Rithmire, and Kellee Tsai write, "The CCP's approach to economic governance became 'securitized,' such that political control over firms and risk management are prioritized over rapid growth."[39] Adopting Maoist language, Xi proclaimed in his 19th Plenum speech, "It doesn't matter whether it is the government, the military, the people, or the schools east, west, north, south, or the center, the Party rules everything."[40] Shirk describes Xi's effort to make indistinguishable the government and the Party, by requiring the heads of all state institutions to routinely report to the Politburo Standing Committee.[41] Since the 2008 financial crisis, the Party has moved to penetrate all of Chinese society, aiming to create Party offices within all organizations. This includes private-sector firms (including foreign firms), NGOs, and universities.[42] "The party has called on private companies to help state enterprises, demanded the rich give back to society through philanthropic giving, and passed a raft of legislative oversight," comments Tony Saich. "These new regulations have gone far beyond seeking to control and regulate the real estate and tech sectors. For example, they've led to disbanding the for-profit education sector and dictating how many hours children can play games online."[43] The CCP has taken direct ownership in private firms and passed a host of regulatory reforms to control the firms' behavior. The United Front Work Department (which also monitors civil society organizations) now meets with business owners to convey CCP expectations and to make sure that businesspeople "identify politically, intellectually and emotionally" with the Party.[44]

Observers argue that the Xi government's efforts to crack down on the technology sector will stifle Chinese innovation in the future. People point to the CCP's smacking of the ride-hailing app DiDi, after the company disobeyed government diktat to delay its initial public offering (IPO). Most famously, the CCP brought the hammer down on Alibaba Group's CEO, the self-made billionaire Jack Ma, abruptly canceling the $37 billion IPO for the Alibaba affiliate Ant Group after a notorious speech in which Ma publicly criticized the government.[45] The superstar CEO was forced into retirement and has faded from prominence. As Li Yuan argued in the *New York Times*, through its crackdown on tech, "Beijing tamed the industry's ambition and blunted its innovative edge."[46]

Although government controls within China's highly innovative technology sector may indeed dampen innovation, such criticism is overstated. Accounts of the tech crackdown frequently neglect the many sound reasons, related to concerns about data protection and the risk of financial crises, why the CCP imposed greater regulatory control. (Indeed, the US government has similar concerns; as one commentator wrote, "China is doing what the U.S. can't seem to: regulate its tech companies.")[47] Furthermore, critics registered the tightening but not the loosening; the CCP has since shifted its policy, ending its probe of Ant Group and reissuing the firm's licenses (which the government had previously withheld) to engage in the gaming industry.[48] Today, in response to economic sanctions from the United States, the CCP "broadcast the government's intention to enlist the tech industry in its broader ambition to counter US efforts."[49] Furthermore, critics ignore interesting sectoral variation; government regulations in the artificial intelligence (AI) realm have been kept deliberately lax in order to encourage innovation.[50]

In this interpretation, Xi's crackdown on the tech sector is not a baffling own goal committed by an uninformed or irrational leader; rather, it adheres to a smart authoritarian logic of allowing freedoms when you can, imposing controls when you must. "Loosening causes chaos; tightening up causes death," Angela Huyue Zhang quotes in her book about the tech crackdown. The saying, she argues, "perfectly captures how the regulators dramatically yet predictably oscillate between doing too little to police the tech sector and doing too much."[51]

Some aspects of the CCP's controls over the technology sector do raise questions about the state's role in Chinese innovation. In the Made in China 2025 plan, the CCP identified several strategic sectors: information technology such as AI and the Internet of Things; robotics; green energy and electric vehicles; aerospace; ocean engineering; railway, agricultural, and power equipment; new materials; and medicine and medical devices. Because of growing political tension and US economic pressure since first the Trump administration, China invested massively in strengthening its indigenous semiconductor manufacturing capabilities. Through its more intrusive policies, the CCP is forcing the private sector to shift out of "soft" technological pursuits into "hard" ones. In other words, for example, the Party doesn't want firms devoting resources toward developing a slightly better food delivery app—it wants them focusing on the sectors that the government sees as the foundation of national power. (Ironically, a similar shift toward industrial policy is occurring in the United States, notably in the semiconductor industry.)[52]

China's shift toward a heavier hand for the state is frequently lamented as another misguided policy. But, as discussed in Chapter 2, government involvement need not quash innovation; indeed, it depends on implementation. As Linda Weiss argued, Washington's funding and leadership played a crucial role in the creation of Silicon Valley and in encouraging the developments of key consumer sectors and military technologies. Importantly, she notes that the US government delegated innovation to universities and private firms.[53] America's "techno-security state," describes Tai Ming Cheung, was "highly pluralistic and decentralized."[54] As Adam Segal argues, the US "software" or micro-institutions that drive innovation—for example, peer review and grant-seeking processes, relationships between universities, government, and financial institutions, and so on—contribute significantly to its success.[55] In other words, state intervention is not necessarily a bad thing—what matters is the way that the government intervenes, and whether China can continue to develop a healthy innovation ecosystem.

THE END OF STATE-SECTOR REFORM

As Chapter 1 described, numerous reforms are essential to shift an economy from catch-up to innovation-based growth. While China has successfully encouraged this shift in many ways—for example, the development of a high level of human capital—the CCP's reform of the state sector is lagging. In 2013 in the Third Plenum, the CCP articulated sweeping reform goals for the state sector and the judiciary. But, notes David Shambaugh, "by virtually all foreign evaluations, only a meager 14 to 20 percent of the Third Plenum package has been implemented."[56] Shirk agrees that "as Xi has taken command of every ligament in the system, he has not used his power to carry out these market reforms: most of the proposals remain on the drawing board."[57] Although the state sector increased its profitability after multiple rounds of reform, its firms still benefit from preferential treatment in terms of government loans and unfair barriers to entry that keep private firms out of some sectors.[58]

The resurgence of China's state sector will dampen Chinese innovation. China's private sector accounts for the majority of Chinese innovation, and for almost all (> 99 percent) of its high-tech exports.[59] The state sector, however, continues to hoover up most of the capital.[60] The sector's debt burden is massive, and its profitability and innovation are lower than the private sector's. If the CCP does not manage to reform the state sector, Chinese innovation will suffer. Perhaps China can still maintain its technological competitiveness—but this will be a particularly challenging area for the CCP to navigate.

RETURN TO PERSONALISM

Scholars also argue that as the Party attempts greater control over the Chinese economy, greater personalism is diminishing the quality of governance. Noting the bad old days of Mao's governance before the reform era, Shirk comments that "the system appears to be reverting to dictatorial rule." Xi, Shirk argues, is "scrapping the institutional norms and precedents that the party has built around its collective leadership since 1979."[61] This includes reducing collective leadership and concentrating power in Xi's hands.[62]

Indeed, several of Xi's policies have led observers to argue that China is headed back to personalism. Xi has overturned term limits and age limits created during the reform era; although his own term had been set to expire in 2022, the 2018 National People's Congress eliminated the two-term limit for the presidency. Kevin Rudd notes that the 20th Party Congress selectively applied term limits to other positions; it

> removed more reform-minded party officials who had sometimes disagreed with Xi, such as Premier Li Keqiang and Wang Yang, from the Politburo Standing Committee, and it removed the reformist Hu Chunhua from the wider Politburo—even though none of them had reached the retirement age of 68. Meanwhile, the congress allowed other political loyalists over the retirement age to stay. (One of them, Zhang Youxia, the vice chairman of the Central Military Commission, is already 72.)[63]

When it comes to term limits, Xi is violating norms (described in Chapter 4) that previous Chinese leaders instituted to promote political stability. The CCP respected these norms for decades; such norms undergirded "two of the most stable transitions of power in China's modern history, from Mr. Jiang to Mr. Hu in 2002, and then Mr. Hu to Mr. Xi."[64] These norms served as a Chinese version of checks and balances, which the CCP under Deng's leadership created to improve the quality of governance and to prevent China from ever again experiencing the personalist rule—and chaos—of the Mao years. In sum, in recent years, Xi has undone many of the reforms instituted under Deng that were aimed at decentralizing and professionalizing the CCP.

Greater personalism could undermine Chinese innovation. If cadres are selected, promoted, and assigned on the basis of political loyalty rather than competence, this violates one of the key attributes of smart authoritarian rule. At a time when the CCP is increasing the level of state involvement in innovation, assigning loyalist toadies to key economic roles in the bureaucracy would be disastrous for innovation and for the economy more broadly. Furthermore, as described earlier, the success of the smart authoritarian model depends on government leaders having reliable information. Although the government censors the external media, the CCP relies on an internal media (*neican*) for a frank presentation of challenges facing the country. If personalism leads analysts to be afraid to present accurate information to Chinese government officials, this could also undermine China's smart authoritarianism—and ultimately innovation.

ELDERLY RUBBISH

Scholars also observe a crackdown on freedoms, civil society, and information flows that they argue will undermine innovation. "Since Xi came to power in 2012," write Xu and Guo, "China has shifted back toward totalitarianism, with the CPC leadership reasserting control, particularly over the burgeoning private sector."[65] First, Xi has cracked down on Chinese civil society. In 2013, CCP leadership issued numerous communiques, the ninth of which ("Document Nine") described the threats China faced from liberal countries. This communique argued that liberal states were trying to weaken the Party by empowering civil society: "Western embassies, consulates, media operations, and NGOs [nongovernmental organisations] operating inside China under various covers are . . . cultivating . . . anti-government forces."[66] Document Nine led to hundreds of arrests of civil rights lawyers, feminists, and other activists. "These arrests sent a chill through non-profit organisations (NPOs) and caused many to recalibrate their work in light of what appeared to be new limits on permissible activity."[67] In 2016, the Overseas NGO Management Law required NGOs to register with the police (the Ministry of Public Security), suggesting they represented a national security threat.[68]

Government repression has increased in several other areas. Religious worship is subjected to increased state control, with many religious leaders surveilled and arrested. Repression is particularly severe in Xinjiang, the home of the Muslim Uighur minority.[69] Furthermore, the media face significantly more censorship and government control,[70] and reporters and editors who anger the regime are more likely to be arrested than in the previous "golden age" of relative media freedom.[71] Xi visited media outlets in 2016 to inform them that "their work must reflect total loyalty to the party."[72] Furthermore, Xi's government has pursued more extractive policies by intruding more deeply into people's private lives. The government deplored and sought to regulate how many hours of video games children play;[73] Xi condemns "vulgar influencers" and effeminate males;[74] the government has sought to crack down on private tutoring (arguing that it promotes inequality).[75]

Furthermore, Xi has intensified propaganda with "Xi Jinping Thought." At its essence, this doctrine holds that China must once again achieve its position of national greatness, and that only the Chinese Communist Party—and only Xi leading it—can deliver this bright future. Xi Jinping Thought is now everywhere; "Schools, newspapers, television, the internet, billboards and banners all trumpet the ideas of Mr. Xi, the country's president and Communist Party leader."[76] In 2021, school curricula were modified to include Xi Jinping Thought, in order to help "teenagers establish Marxist beliefs."[77] Two university departments were established for the study of Xi Jinping Thought, in part for the purpose of creating university curricula.[78] In W. H. Auden's *Another Time* (which laments the rise of fascism), the poet disdains authoritarian propaganda as "elderly rubbish." At the start of the twenty-first century, when the Chinese people have more wealth and technology to nurture their creativity than ever before, the CCP is feeding them elderly rubbish.

CHINA'S FUTURE

In sum, the skeptical view about China's future attributes its unexpected success in innovation to the loosening of restrictions around the turn of the twenty-first century. But (as expected by institutions theory and modernization theory) this unleashed societal forces that challenged CCP control and led the Party to clamp down. Xi's abandonment of state-sector reforms, increased personalism, and so forth will quash innovation, leading China to follow the Soviet Union's slide off the global technology frontier.[79] In this view, China's smart authoritarianism and innovation success were an unexpected interregnum, but in the long term, authoritarianism—even the "smart" kind—is incompatible with innovation-based growth. This view may be proven correct.

A few caveats before liberalism can declare victory. First, in defiance of institutions theory, authoritarian regimes can clearly generate innovation, even if only for a while. But even if smart authoritarianism is just an interlude, in Singapore's case it's been quite a long interlude—it's currently on its sixth decade. In China's case, if smart authoritarianism lasts only about 50 years, a lot has happened during those years: China returned to the great-power ranks, shifted the international system from unipolarity into bipolarity, and dramatically transformed U.S. national security strategy.[80] It engaged the United States in a great-power rivalry in which China may eventually come to establish regional hegemony in East Asia.[81] Through its relationship with Russia and its support of autocratic regimes, China has supported a trend in democratic reversals around the world.[82] So even if China's smart authoritarianism is just an interregnum, it has been a long and consequential one.

Second, it may not be an interregnum; the CCP may pull this off. In the smart authoritarian view, the Party's move toward greater control, which many observers interpret as bafflingly misguided, was a tactical response to years of more relaxed policies. Indeed, an ongoing topic among Party leaders and intellectuals is the balance between democracy and authoritarianism, and the need—after periods of reform—for a "neo-authoritarian" turn to correct for problems that arise during liberalization.[83] Thus, scholar Xiao Gong Qing characterizes Xi's policy shifts not as regression to Mao but as a prudent response to liberalization. "If Deng Xiaoping developed the 1.0 version of neo-authoritarianism," Xiao argues, "Xi is now developing the 2.0 version. Xi wants to use an enhanced version of neo-authoritarianism to achieve Deng's stated goals."[84] In sum, the smart authoritarian view holds that the CCP's neo-authoritarian turn was calculated and necessary; it will enable the Party to retain power while still permitting enough innovation and growth for China to compete effectively. All depends on the CCP's continued management of the tensions in the smart authoritarian model—that is, its ability to tack between more freedoms and more control.

China could fail at smart authoritarianism in one of two ways. First, as many experts are predicting, the CCP might tack too far toward stultifying control: infecting government control and corruption into what had been thriving and innovative sectors, and isolating China from key markets, technology partners, and research and educational networks. The second failure mode is *perestroika*: the result of tacking too far toward freedom and openness, with the CCP losing power as a result. Either one of these outcomes would represent a failure of smart authoritarianism.

Some scholars may argue that herein lies the democratic advantage: that autocrats face inherent tensions between their ability to stay in power and their ability to compete in a modern, globalized economy. But as described in Chapter 3, democracies struggle with tensions too: between minimizing taxation and government interference for the purpose of encouraging innovation, entrepreneurship and maximizing economic growth, as opposed to enacting robust government intervention in the economy for the purpose of regulating harmful business activities, accounting for externalities, and redistributing wealth. Due to globalization and technological change, these tensions have caused significant domestic political instability in liberal countries, including the United States. As liberal states and authoritarian states continue to intervene in each other's societies in an attempt to weaken one another, liberal states also face vulnerabilities from their own openness, which facilitates intervention by authoritarian rivals.

In one sense, the CCP is grappling with the age-old "king's dilemma" of trying to balance the need for control against the need for reforms. In another sense, however, the CCP finds itself in uncharted territory. "People have called it market Leninism, authoritarian capitalism. . . . We are watching a kind of petri dish in which an experiment of extraordinary importance to the world is being carried out," observes Orville Schell. "Whether you can bring together a one-party state with an innovative sector—both economically and technologically innovative—that's something we thought could not coexist."[85] First, the nature of innovation in the Fourth Industrial Revolution (the 4IR) is uncharted territory; it's uncertain how smart authoritarianism will affect innovation.[86] Arguments that Xi's authoritarianism will stifle innovation rest on a view of innovation as homogenous and static, when in fact it is heterogenous (comprised, as Chapter 2 described, of many different types) and dynamic (as the nature of innovation changes with the new technologies of the 4IR, particularly AI). It's possible that China's smart authoritarianism will dampen some types of innovation but not others.[87] It's also possible that in the 4IR, the previous tension between innovation and authoritarian control will no longer exist.[88]

We are also in uncharted territory with respect to the methods used by autocratic regimes to stay in power. The CCP fields unprecedented tools of control, relying on facial recognition data analyzed by AI, on spyware installed by governments on cell phones to track dissidents at home and exiles abroad, and increasingly on the exploitation of wearable technologies and the Internet of Things (e.g., biometric data, medical monitors, and emotional detection technology).[89] "Digital authoritarians" may be able to exert significant control in ways that do not undermine innovation. Although the future is unclear, this book, using the lens of smart authoritarianism, explains how China got where it is today: to a place many observers said it couldn't reach.

The Implications (and Nonimplications) of China's Rise

Many observers of China's rise have argued that although China has the scale of a great power, it lacks the requisite technological sophistication. Facing a large gap vis-à-vis the United States, and hamstrung by its authoritarian model, such skeptics have argued, China would be unable to compete against the United States in great-power politics. Elsewhere I have shown that China easily exceeds the threshold for great power (and indeed according to some metrics dramatically exceeds it); in other words, China is a great power and superpower whose rise has returned the international system to bipolarity.[90] This book has demonstrated that China is also successfully competing at the global technological frontier. Thus China, through its economic, technological, and military rise, has become a peer competitor of the United States.

China's rise to great power—which challenges beliefs about a democratic advantage in great-power competition—can be explained by smart authoritarianism. Institutions theorists were correct that autocrats found themselves at a disadvantage in the late twentieth century because of the information age, the liberal development regime, and other trends. But observers erred in seeing authoritarians as homogenous and static. Some authoritarians adapted: they shifted to a "smart" authoritarian model more conducive to economic success. They accepted greater institutional constraints on their rule, tempered repression, and adopted more inclusive economic policies.

Smart authoritarianism relies on a regime's calibrations between more control and more freedoms. When leaders fear that their political control is slipping, they will tactically rein in freedoms; at other times, regimes will allow greater freedoms in order to encourage growth and innovation. Some regimes fail to find this balance and are ousted—as shown by the experiences of the shah of Iran and former Soviet leader Mikhail Gorbachev. Other authoritarian regimes (South Korea and Taiwan) liberalized. But some smart authoritarians stay in power a long time, and—in defiance of skeptics—can cultivate economic growth and innovation. As the Chinese case shows, this can have powerful consequences for international politics.

IMPLICATIONS

Findings from this book contribute to key policy debates and inform multiple scholarly literatures. First, China has not only the scale but also the technological sophistication of a great power—in fact, a superpower.[91] China is likely to behave as superpowers normally behave. That is, it will likely seek regional hegemony: undermining US alliances, influencing (overtly and covertly) its neighbors' political systems to encourage pro-China policies, and attempting to control territories that China lost in the nineteenth century.[92] China will try to influence international norms and technology standards in ways that favor its own firms and political system.[93]

Second, because of a Chinese superpower, the United States—rather than enjoying continued unipolarity and the "freedom to roam" that comes with it—faces a far more constrained and dangerous international environment than it has for the past 30 years.[94] Superpower China is already seeking close and multidimensional ties with governments in the Persian Gulf, who can play Beijing and Washington off one another. China's influence has also grown in Latin America; as during the Cold War, that region will likely become a focus of superpower contestation (only this time with a significant economic dimension as well).[95] In these emerging geopolitical conditions, the United States may need to defer to a rival's sphere of influence and—to balance against a more formidable superpower than the Soviet Union ever was—ask more of its allies, as was necessary during the Cold War.

This book also informs the conversation about the intensifying global struggle between democracy and authoritarianism. "Democracy is in real danger all over the globe," argues Michael J. Abramowitz, president of Freedom House. The NGO reports a 16-year decline in countries' freedom scores.[96] Indeed, observers are watching with alarm democratic backsliding in, for example, Georgia, Hungary, Mali, the Philippines, and Zimbabwe. Leaders there and in other countries are increasingly taking greater control of the media, making elections less competitive, and in other ways reducing the quality of democracy.[97] Even in India and the United States—large, prominent democracies—many observers warn of democratic erosion. Freedom House argues that "The leaders of China, Russia, and other dictatorships have succeeded in challenging the consensus that democracy is the only viable path to prosperity and security."[98]

Smart authoritarianism suggests that democracy faces an even graver threat than presently believed.[99] Under this model, dictators are more resilient and more geopolitically competitive than institutions theories expected. To use epidemiological terms, the delta variant of authoritarianism has shifted to the more competitive omicron variant. Today's smart authoritarians also have new and powerful offensive tools at their disposal that they are wielding globally to undermine democratic rivals.[100]

THEORETICAL IMPLICATIONS

Findings from this book also contribute to debates within political science and social science more broadly. The book challenges a highly prominent multidisciplinary literature about the relationship between extractive institutions and innovation-based growth. Future work should push past the democratic/authoritarian binary to explore heterogeneity among authoritarian regimes; additionally, as scholars debate how authoritarian rule affects innovation, scholars should explore different types of innovation.

Furthermore, this book builds upon, and contributes to, a growing literature on authoritarian politics that explores heterogeneity among authoritarian regimes. Standing on the shoulders of a large comparative political economy literature on developmental authoritarianism, this book shows that not only can authoritarian countries preside over catch-up growth, they can also cultivate cutting-edge innovation. This has profound implications for how we view the reformist regimes in the Middle East, Central Asia, and elsewhere.

Furthermore, this book contributes to an IR literature on "democratic difference," which expects liberal countries to compete more effectively than illiberal states in a variety of ways. This book (and indeed the large authoritarian politics literature on authoritarian heterogeneity) underscores the need for IR scholars to revisit theories of democratic difference. IR theorists have only just begun to push past the democratic/authoritarian binary to take into account authoritarian heterogeneity.[101] This book contributes to and hopefully accelerates this trend.

NONIMPLICATIONS

It's also important to clarify what this book is **not** arguing. First, I argue that smart authoritarians perform better than was previously believed—

not that they perform better than democracies in general.[102] In other words, I argue neither for superiority of smart authoritarianism generally, nor for superiority of Chinese authoritarianism specifically. Rather, I argue that many IR scholars have neglected to observe significant heterogeneity among authoritarian regimes, which has led scholars to underestimate Chinese power. An overarching theme of the book is that regime type is a poor lens for understanding a country's ability to perform economically and militarily: that state power stems from the policies pursued by regimes of various types.

Second, this book is not a story of US decline. The shift to bipolarity stems from China's rise, not the United States' decline (in other words, from relative rather than absolute decline).[103] Americans love to worry about decline; current debates about the United States losing its technological lead[104] are only the latest in several rounds about the country's diminishing power.[105] Of course, as Amar Bhidé memorably put it, "The United States can't count on the same ending to every episode of the Losing Our Lead serial."[106] But the United States, with all its problems, remains the world's most powerful country and, as many scholars have described, enjoys tremendous competitive advantages.[107]

Unlike most industrialized countries and unlike China, the United States enjoys favorable demographics (stemming from immigration). Chapter 2 showed that the United States maintains a technological lead, and that—given current US performance in emerging technologies—its strong performance will likely continue.[108] Washington's shift toward industrial policy, aimed at sustaining US technology leadership vis-à-vis China, will benefit from well-developed microinstitutions that encourage innovation.[109] The United States enjoys close relations with the world's advanced industrialized economies, and also has significant soft power resources. Washington is relying on its global influence to "weaponize" interdependence against China, which (as discussed in Chapter 2) is complicating China's technological rise.[110] Just as it's important to understand the extent of China's challenge, it's also important to understand the United States's formidable national power and many advantages in science and technology (S&T).

Third and relatedly, this book offers no predictions about the outcome of a US-China military confrontation. I argue that China is a peer competitor in terms of both scale and sophistication, and as a great power, can "put up a serious fight" against the United States.[111] But great powers vary significantly in their battlefield effectiveness.[112]

Whereas the US military fights with tremendous effectiveness, the People's Liberation Army (PLA) remains untested, its battlefield effectiveness very much in question.[113] Indeed, historically the CCP pursued defense policies (e.g., coup-proofing, corruption, low human capital) that typically undercut battlefield effectiveness.[114] Since 2014, however, Xi has presided over a round of important military reforms. The CCP seeks to streamline its notoriously bloated defense sector, to make military promotions on the basis of competence rather than corruption, to increase the realism of military training, and to improve jointness and horizontal communications in the PLA's formerly stove-piped institutional setting. [115] Whereas such reforms suggest promise for future PLA military effectiveness, their success remains uncertain. Russia's invasion of Ukraine provides a glaring example of an authoritarian military that (based on recent modernization) appeared strong on paper but performed poorly on the battlefield.[116] Arguments in this book offer little insight about the outcome of a potential US-China war; they only suggest that as a great power with large and technologically advanced military capabilities, China has put itself in the game.

LOOKING AHEAD

This book's identification of China as a superpower, and the phenomenon of smart authoritarianism, raise important questions for future research by political scientists, innovation scholars, economists, and others.

Competing with Smart Authoritarians. First, the claim that autocracies can be far tougher competitors than assumed by many scholars suggests the need to update the debate about the struggle between democracy and authoritarianism. Prodemocracy activists and researchers cannot afford the view that authoritarian regimes are inherently disadvantaged as they compete with democracies. Authoritarians have reinvented themselves and their tools of control, and in many places are thriving. Researchers should explore how democracies can best compete against this new breed of authoritarians: where their vulnerabilities lie, and in which ways democracies are the most vulnerable to their intervention and manipulation.[117]

Authoritarian versus Democratic Innovation. Second, this book encourages a research program about innovation in democratic versus authoritarian societies. Many observers expected that China's authoritarian institutions would prevent it from successfully transitioning to innovation-based growth; this book shows such pessimism to be misplaced. Clearly, authoritarian countries can indeed innovate under some conditions, as seen in China and Singapore, as well as preliberalization South Korea and Taiwan.[118] Thus scholars from quite different subfields—for example, comparative politics and innovation—should research the relative strengths of different types of regimes in producing advanced technology.

For example, perhaps democracies and authoritarian regimes have different advantages that lead them to be better at producing some types of innovation (e.g., science-based, customer-focused) relative to others. Drawing on arguments in Charles Lindblom's *Politics and Markets* about economic growth in different types of political-economic systems, observers speculate that perhaps autocracies (with strong central governments) should excel at innovation that requires central leadership and large expenditures: "moonshot" projects and basic research. Democracies would be expected to perform relatively worse at such activities because they have less centralization, shorter time horizons, and excessive competition.

The cases of the United States and China, however, suggest the picture is more complex than this; the democratic United States is a world leader in basic research, with a long history of government involvement in technology. China performs well in customer-focused, grassroots innovation, as in its financial technology (or "fintech") sector. (Furthermore, with dramatic regional variation, and with the decentralization of economic policy to provincial and local levels during the reform era, China is more decentralized than many observers appreciate.) In short, the picture is clearly more complex than this first-cut hypothesis would lead us to expect. Importantly, this question arises only after we internalize that authoritarian countries are heterogenous, and that some of them can produce cutting-edge innovation. This fact highlights new questions for future researchers to explore.

Innovation in the 4IR. The evergreen debate about the relative economic performance of democracies and autocracies takes on new significance as the world moves into the 4IR. Among the many emerging technologies identified as fueling future productivity increases and military power, at this point AI stands out as the most significant.[119] Already scholars are speculating whether democracies or autocracies are better positioned to capture significant productivity gains from the AI revolution. Some scholars argue that autocracies' investments in AI for domestic political control purposes support the society's adoption of the technology and thus boost its global AI competitiveness.[120] The fact that tens (hundreds?) of thousands of Chinese local- and provincial-level officials are trained in AI methods suggests this is one way the technology will diffuse throughout China's economy. Furthermore, AI models require vast amounts of data, which autocrats possess due to their use of AI for political control. Other observers, however, argue that far from "data being the new oil," the data required for training AI models are highly specific.[121] Future research should build on this debate and build on the arguments within this book about authoritarian versus democratic innovation in AI and in the other technologies of the 4IR.

Social scientists indeed have many questions to investigate about the evolution of authoritarian regimes, which is occurring amid profound technological change. Meanwhile, this book shows that China—underestimated by many observers—not only helped reinvent authoritarianism but, by becoming a global technology leader, transformed the global balance of power.

Notes

Introduction

1. On polarity, see Lind 2024; Tunsjø 2018; Monteiro 2014; Mearsheimer 2014; Posen 2011. On power transitions and conflict, see Gilpin 1981, 158; Allison 2017; Shifrinson 2018; Schake 2017; Kliman 2015; Organski and Kugler 1980.

2. On war, see Allison 2017; Mastro 2021. On international order, see Doshi 2021; Rolland 2020; Mazarr, Heath, and Cevallos 2018.

3. Dukalskis 2021; Gokhale 2020; Kendall-Taylor and Shullman 2018; Dobson 2013; Brady 2009.

4. On China's rise and the future of US grand strategy, see Allison 2020; Lind and Press 2020; Lind and Wohlforth 2019; Wertheim 2020; Brooks, Ikenberry, and Wohlforth 2012. On changes in Japanese national security policy, for example, see Lind 2022; Liff 2015.

5. Brooks and Wohlforth 2023; Gilli and Gilli 2019; Beckley 2018; Brooks and Wohlforth 2016; Christensen 2015.

6. Brooks and Wohlforth 2015, 34.

7. Gilli and Gilli 2019, 156.

8. On "peak China," see Beckley and Brands 2022; Beckley and Brands 2021. On China's growth challenges, see Vagle and Brooks 2025; Shambaugh 2016.

9. On Xi's leadership, see Shirk 2022; Shambaugh 2021; Lardy 2019; Minzner 2018a; Economy 2018; Magnus 2018.

10. On a "middle income trap," see Fuller 2016; Aiyar et al. 2013; Kharas and Kohli 2011. On the requirements of catch-up vs cutting edge growth see Cowen and Tabarrok 2015.

11. On institutions and growth, see Acemoglu and Robinson 2012; Acemoglu 2008; Pei 2006; Acemoglu, Johnson, and Robinson 2005; North 1990; North, Wallis, and Weingast 2012.

12. Packard 2010, 86.

13. Atkinson 2024; Allison et al. 2021; Kai-Fu Lee 2018. Also see Toner et al 2023.

14. On China's innovation system, see A. B. Kennedy 2019; Segal 2011; Ji 2023. On more competitive relations with the United States and Europe, see Swanson, McCabe, and Crowley 2023; Liu and Chang 2023; Duchâtel 2021; M. Y. Zhang 2023.

15. Morgenbesser 2020; Kerkvliet 2014; Kendall-Taylor and Frantz 2014; J. Wright 2008; Policzer 2009; Geddes 1999; Krastev 2011.

16. Guriev and Treisman 2022; Dukalskis 2021; Morgenbesser 2020; Kerkvliet 2014; Kendall-Taylor and Frantz 2014b; Dobson 2013; Krastev 2011; Policzer 2009.

17. Lind 2024.

18. Doshi 2021; Mearsheimer 2014. On Chinese influence operations and domestic political intervention, see Ohlin and Hollis 2021; Lind 2018; Brady 2009.

19. On Chinese efforts to shape international order, see Rühlig 2022; Oud 2020; Rolland 2020; Mazarr, Heath, and Cevallos 2018; Brunnermeier, Doshi, and James 2018.

20. On exporting authoritarianism, see Dukalskis 2021; Kendall-Taylor and Shullman 2018; Polyakova and Meserole 2019. On the struggle between democracy and authoritarianism, see Levitsky and Way 2010; Levitsky and Ziblatt 2018; Guriev and Treisman 2022.

21. Way 2023; Oud 2020; Gokhale 2020.

22. Beckley and Brands 2021, 2022.

23. Shirk 2022; Lardy 2019; Magnus 2018; Shirk 2018; Minzner 2018; Shambaugh 2021.

24. A. H. Zhang 2024.

25. On different types of innovation, see Atkinson and Foote 2019; Woertzel et al. 2015; Pavitt 1984.

26. Chin and Lin 2022; Xu Xu 2021; Roberts 2018.

27. Beraja et al. 2023.

28. See, for example, R. Brooks 2008; Weeks 2014. Security studies scholars have begun to explore the heterogeneity of autocratic policies of coup-proofing; see Talmadge 2015; Reiter 2020.

29. Taylor 2016; Doner and Schneider 2016.

30. Discussing mechanisms of a democratic advantage are Drezner 2022; Hyde and Saunders 2020; Kroenig 2020; Baum and Lake 2003. On conditions to support innovation, see Taylor 2016.

31. Guriev and Treisman 2022, 2019; Frantz 2018; Glasius 2018b; Kerkvliet 2014; Policzer 2009; Morgenbesser 2020; Krastev 2011.

32. Shambaugh 2016; Beckley and Brands 2021. On China's "zero-COVID" policy, see Ang 2022.

33. Shirk 2022; Shambaugh 2021; Lardy 2019; Minzner 2018; Economy 2018; Magnus 2018.

1. The Stakes and the Debate

1. Spufford 2012.

2. Ikenberry 2000; Ikenberry 2012; Lascurettes 2020; Rolland 2020.

3. Friedberg 2015; Allison 2017; Mearsheimer 2014; Copeland 2000.

4. Lind and Shifrinson 2020; on subversion, see Wohlforth 2020.

5. O'Rourke 2018; Carson 2018.

6. On power transition, see Shifrinson 2018; Allison 2017; Schake 2017; Kliman 2015; Blainey 1988; P. Kennedy 1987; J. S. Levy 1983; Gilpin 1981; Organski and Kugler 1980.

7. This book follows scholars who define great powers as having a set of capabilities or resources, as opposed to a state's ability to get another state to behave in a desired manner. For discussion, see Beckley 2018a; Beckley 2018b; Mearsheimer 2014; Monteiro 2014; Tellis et al. 2000; Levy 1983.

8. Mearsheimer 2014; Beckley 2018b; Shifrinson 2018; Monteiro 2014; Levy 1983.

9. Levy 1983, 44; Mearsheimer 2014; Beckley 2018b; Shifrinson 2018; Monteiro 2014.

10. On soft power and status, see Nye 2004; Murray 2018; Paul, Larson, and Wohlforth 2014.

11. Ding 2024, 2021; Brooks and Wohlforth 2015; Beckley 2011; Horowitz 2010.

12. Brooks and Wohlforth 2015, 16.

13. Rasler and Thompson 1994, 7; Thompson 1990; Ding 2021.

14. Brodie and Brodie 1973, 119.

15. Debating the causes of the Industrial Revolution are Pomeranz 2000; Landes 1969; Mokyr 2016; Frank 1998; Hobson 2004.

16. W. A. Murray 2005, 223–24.

17. Brodie and Brodie 1973, 157.
18. W. A. Murray 2005; Worthing 2007.
19. Perry 2004.
20. Boot 2006.
21. Worthing 2007, 36; M. C. Wright 1957; De Bary and Lufrano 1999.
22. Worthing 2007, 36.
23. Brodie and Brodie 1973, 137.
24. Boot 2006, 291.
25. Boot 2006, 210.
26. Brooks and Wohlforth 2000; Galambos 2012; Cortada 2012.
27. Cohen 1996, 39.
28. Terminology from Cowen and Tabarrok 2015.
29. Dyson 2013; Bloom, Canning, and Sevilla 2003; Bloom and Williamson 1998.
30. Cowen and Tabarrok 2015; Krugman 1994.
31. See, for example, Lamb 1982.
32. Pritchett and Summers 2014; Sharma 2012; Young 1995; Krugman 1994.
33. Bloom, Canning, and Sevilla 2003; Bloom and Williamson 1998; Dyson 2013.
34. Cowen and Tabarrok 2015.
35. Krugman 1994.
36. Eichengreen, Park, and Shin 2013; Kharas and Kohli 2011; Foxley and Sossdorf 2011.
37. Taylor 2016, 29.
38. Taylor 2016, 29. See his Appendix 1A for a superb discussion of how to define *innovation*.
39. OECD 2015.
40. Kharas and Kohli 2011, 285.
41. Kharas and Kohli 2011, 285.
42. Lind 2024.
43. For a brilliant summary, see Taylor 2016.
44. Arrow 1962; for discussion, see Breznitz and Murphree 2011; Taylor 2016.
45. Patents are not the only method; scholars note that IPR confers a monopoly rent, and they describe other incentives (such as prizes) for innovation. For discussion, see Boldrin and Levine 2013; Milgrom and Roberts 1992; Schumpeter 1934. For example, under the 1714 Longitude Act, the British government offered a prize (worth over $3 million today) to the person who could solve a critical problem of ship navigation; the prize was awarded to John Harrison for his invention of the marine chronometer. In the Soviet Union, the government encouraged innovation not through patenting but by rewarding inventors with profit shares, large bonuses and other perquisites, and "inventor's certificates" (Soltysinski 1969).
46. Keefer and Knack 1997.
47. Acemoglu et al. 2005.
48. Schumpeter 1911.
49. Rajan and Zingales 1998; Kortum and Lerner 2000; Samila and Sorenson 2011.
50. "The Future of Productivity" 2015, 13.
51. Gill and Kharas 2007; "The Future of Productivity" 2015; Kharas and Kohli 2011.
52. Kharas and Kohli 2011, 287. Tertiary education includes postsecondary education, such as at universities, technical institutes, two-year colleges, and research laboratories.
53. Ding 2021; Skinner and Staiger 2007.
54. White 1994, 377.
55. Ingram 2020.
56. Kirby 2022.
57. Taylor 2016; Breznitz 2007.
58. On such connections, see Taylor 2016; O'Mara 2015; Weiss 2014; Segal 2011.
59. Segal 2011; Porter 1990. For a literature review, see Taylor 2016.
60. Taylor 2016, 141.
61. Taylor 2016, 165–72.
62. World Bank Group 2020, 25.

63. For international engagement as a sign of university quality, see Kirby 2022, 15. On human capital and the "globalization of innovation," see Kennedy 2018.
64. Armstrong 1984.
65. Aiyar et al. 2013.
66. Olson 1982. See also Rodrik 1992; Hirschman 1991. On this dynamic with respect to innovation, see Taylor 2016.
67. Foxley and Sossdorf 2011; Doner and Schneider 2016; Waterbury 1992; Caballero, Hoshi, and Kashyap 2008. On education reforms, see Grindle 2004.
68. For an excellent summary, see Kroenig 2020.
69. North 1990.
70. North 1990.
71. Rodrik 2007, 157.
72. Acemoglu and Robinson 2012; North, Wallis, and Weingast 2012; North 1990; Olson 1982.
73. Acemoglu and Robinson 2012, 75–77.
74. Acemoglu and Robinson 2012, 76.
75. Acemoglu et al. 2008.
76. On mechanisms, see Drezner 2022; Hyde and Saunders 2020.
77. Mkandawire 2015; X. Lu 2000; Wedeman 1997.
78. Mesquita et al. 2003; Baum and Lake 2003. On how education threatens autocratic rule, see Sanborn and Thyne 2014.
79. On debt, see North and Weingast 1989. On encouraging investment, see Barro 1996.
80. On patronage, see Mkandawire 2015; Lienert and Modi 1997; Van de Walle 2001. On defaulting on debt, see Arias, Hollyer, and Rosendorff 2018, 907; also see Gehlbach and Keefer 2011, 2012;; Fang 2011. Asserting the inability of autocracies to make binding agreements are Acemoglu et al. 2008. On interfering with monetary policy, see Rogoff 1985; Alesina and Summers 1993. On illiberal autocratic trade policy, see Bhagwati 1978; Shatz and Tarr 2000; Todaro 1977.
81. Olson 2000, xiii. On patronage, see Mkandawire 2015; Lienert and Modi 1997; Van de Walle 2001. On institutions and credible commitments that encourage investment, see North and Weingast 1989; S. G. Brooks 2005, 57–60.
82. Giuliano, Mishra, and Spilimbergo 2013.
83. Olson 1982.
84. Acemoglu et al. 2019, 51; also see Acemoglu et al. 2008.
85. Imai, Kim, and Wang 2022; Gerring et al. 2005; Przeworski, Alvarez, and Cheibub 2000.
86. Paglayan 2021; Truex 2017; Ross 2006; Mulligan, Gil, and Sala-i-Martin 2004. On incentives for autocrats to invest in public goods, see Gallagher and Hanson 2013; Olson 1993.
87. Kono 2015; Hankla and Kuthy 2013; Dai 2002.
88. Beaulieu, Cox, and Saiegh 2012; Archer, Biglaiser, and DeRouen 2007; Saiegh 2005.
89. Schamis 1999; Haggard and Kaufman 1992, 1995; Evans 1992; Waterbury 1992; Haggard 1990.
90. Haggard 1990; Wade 1990; Kang 2002; Amsden 2001; Young 1995; Krugman 1994.
91. Ang 2020; Kang 2002; Wedeman 1997.
92. Bellows 1993, 145.
93. Popper 2020.
94. Bush 1945.
95. Reif 2021.
96. Cherry 2010; Ingram 2020; White 1994, 377. On public deliberation and growth, see Chandra and Rudra 2015.
97. Gerschewski 2013; Davenport 2007.
98. Quinlivan 1999; Byman and Lind 2010.
99. Byman and Lind 2010.
100. Inglehart and Welzel 2009, 38.
101. Acemoglu and Robinson 2019, 114.
102. Kennedy 2018; Taylor 2016; Doner 2009; Breznitz 2007.
103. Shearer 2004; Matthews 1993.
104. Bremmer 2006, 265.
105. Ding 2021; Mokyr 1992.

106. Peattie 2013; Horowitz 2010; Boyd and Yoshida 1995. On Germany, see Murray and Millett 1998; Mitcham 2008.

107. L. R. Graham 1998; Allen 2001; L. Graham 1990; Ichikawa 2020.

108. Khrushchev 1974, 54.

109. Galambos 2012; Popper 2020; Cortada 2012; Bellows 1993; Daniel Bell 1976.

110. On changes in FDI flows over time, see Trakman and Ranieri 2013. On the emerging significance of FDI and credible commitments, see S. G. Brooks 2005.

111. On the emergence of this aid regime, see Poats 1985; Lumsdaine 1993. On the earlier legacies of this regime, see Martin 2022. On the liberal values of this regime, see Lebovic and Voeten 2009; Kendall-Taylor and Frantz 2014; Schueth 2015.

112. Cooley and Snyder 2015; Marinov and Goemans 2014.

113. Foote 2000; on the South African case, see Klotz 2018.

114. Levitsky and Way 2002, 62.

115. Escribà-Folch and Wright 2015.

116. Huntington 1968.

117. Acemoglu and Robinson 2012, 93.

118. Way 2023.

119. Westad 2015, 454.

120. Brown 2017.

121. Taylor 2016, 127.

122. Taylor 2016, 17.

123. Brummer 2020, 3. See also Breznitz 2007; Doner 2009.

2. The Puzzle of Chinese Innovation

1. Quoted in Liao 1997, 53.

2. Vagle and Brooks 2025; Beckley 2011; Brooks and Wohlforth 2015; Brooks and Wohlforth 2016; Gilli and Gilli 2019.

3. Ellen Ioanes, "China Steals US Designs for New Weapons, and It's Getting Away with 'the Greatest Intellectual Property Theft in Human History,'" *Business Insider*, September 24, 2019.

4. C. Custer, "Why One of China's Most Successful Founders Isn't Afraid to be a Copycat," *TechinAsia.com*, August 31, 2015, https://www.techinasia.com/chinas-successful-founders -afraid-copycat.

5. Will Heilpern, "17 of the Most Shameless Chinese Rip-offs of Western Brands," *Business Insider*, May 6, 2016.

6. For discussion, see Sheehan 2022; Dychtwald 2021; Alexandra Harney, "China's Copycat Culture," *New York Times*, October 31, 2011.

7. Acemoglu and Robinson 2019; Acemoglu and Robinson 2012; Acemoglu, Johnson, and Robinson 2005; Acemoglu et al. 2019; North 1990; North and Thomas 1973.

8. "China Has Become a Scientific Superpower," *The Economist*, June 12, 2024.

9. Schmidt and Bajraktari 2022; Katrina Manson, "US Has Already Lost AI Fight to China, Says Ex-Pentagon Software Chief," *Financial Times*, October 10, 2021; Clay and Atkinson 2022; Wang 2023; Sheehan 2023; Rühlig 2023.

10. Allison et al. 2021, 2. On China's leadership in emerging technologies, see Kaoru Takatsuki, "China Leads High-Tech Research in 80% of Critical Fields: Report," *Nikkei Asia*, September 15, 2023; Kyle Chan, "In the Future, China Will Be Dominant. The U.S. Will Be Irrelevant," *New York Times*, May 19, 2025, https://www.nytimes.com/2025/05/19/opinion/ china-us-trade-tariffs.html; Joe Leahy, Nian Liu, and Ryan McMorrow, "The Lessons from China's Dominance in Manufacturing," *Financial Times*, May 28, 2025, https://www.ft.com/ content/724431ad-26db-4f6d-acab-ccb3cad11daa.

11. U.S. Department of Defense 2021; see also Mastro 2024; Colby 2021.

12. Work and Grant 2019.

13. Ichikawa 2020; Miller 2016; S. G. Brooks 2005; Hanson 2003; Allen 2001; Brooks and Wohlforth 2000.

14. Brooks and Wohlforth 2015, 2016; Beckley 2011, 2018b; Brands 2018; Kroenig 2020.

15. Dan Wang 2023; Woertzel et al. 2015.

16. Taylor 2016, 29. See his Appendix 1A for a superb discussion of how to define innovation.

17. Fagerberg 2006, 4; Breznitz and Murphree 2011; Schumpeter 1934.

18. Schumpeter 1934; Fagenberg 2006, 6.

19. Kline and Rosenberg 1986, 282–84.

20. Breznitz and Murphee 2011.

21. Kline and Rosenberg 1986, 283.

22. Mokyr 1992.

23. Mokyr 1992, 118–19.

24. Bhidé 2008, 6.

25. Bhidé 2008, 6

26. Pavitt 1984, 362.

27. Keith Bradsher, "How China Built Tech Prowess: Chemistry Classes and Research Labs." *New York Times*, August 9, 2024.

28. These innovation categories and descriptions are drawn from Pavitt 1984; Woertzel et al. 2015, 29–31.

29. Pavitt 1984.

30. Woertzel et al. 2015.

31. Di Stefano, Gambardella, and Verona 2012; Woertzel et al. 2015.

32. For discussion, see Bhidé 2008, 10. On debates about measuring innovation, see Harhoff, Narin, Scherer, and Vopel 1999; Eaton and Kortum 1999 Jaffe, Trajtenberg, and Fogarty 2000; Hall, Jaffe, and Trajtenberg 2005.

33. See, for example, Brooks and Wohforth 2015, 22. For useful discussions of innovation metrics, see Taylor 2016; Smith 2006. For superb discussions, see Ding 2021; Taylor 2016.

34. Higher levels of human capital are associated both with invention activities and with diffusion and commercialization of inventions throughout the economy. See Ding 2023; Skinner and Staiger 2007.

35. For discussion, see Taylor 2016, 94–99.

36. Zwetsloot et al., 2021.

37. See Taylor 2016; Brummer 2020.

38. Data from World Intellectual Property Organization, https://www.wipo.int/en/ipfactsandfigures/patents, accessed March 13, 2025. For discussion, see Brummer 2020.

39. Ang et al. 2023; Dang and Motohashi 2015;.

40. Dernis and Khan 2004. Also see Brooks and Wohlforth 2015; Beckley 2011.

41. On the latter, for discussion see Taylor 2016.

42. Skinner and Staiger 2007.

43. Cheung 2022; A. B. Kennedy 2019; Shambaugh 2016; Fuller 2016; Kania 2021b.

44. A. B. Kennedy 2019.

45. Gewirtz 2022, 132.

46. Gewirtz 2022, 203.

47. The National Medium- and Long-Term Program for Science and Technology Development (2006–2020), https://www.itu.int/en/ITU-D/Cybersecurity/Documents/National_Strategies_Repository/China_2006.pdf.

48. Quoted in Cheung 2022, 22.

49. Cheung 2022; Zenglein and Holzmann 2019.

50. Kania and Segal 2021, 311.

51. Sam Snead, "China's Spending on Research and Development Hits a Record $378 Billion," CNBC.com, March 1, 2021, https://www.cnbc.com/2021/03/01/chinas-spending-on-rd-hits-a-record-378-billion.html.

52. Cheung 2022, 21–23.

53. "Full Text of Xi Jinping's Report at 19th CPC National Congress," November 4, 2017, https://www.chinadaily.com.cn/china/19thcpcnationalcongress/2017–11/04/content_34115212.htm.

54. "Xi Says Innovation to Remain at Heart of China's Modernization Drive," Bloomberg Markets and Finance. Xi Jinping Speech to 20th Party Congress, via CCTV. Accessed via Bloomberg Financial News, https://www.youtube.com/watch?v=uz58FQvzGgc.

55. Smriti Mallapaty, "What Xi Jinping's Third Term Means for Science." *Nature*, October 27, 2022, 20–21.

56. Cai 2024

57. Rühlig 2022; Pop, Hua, and Michaels 2021; Doshi 2021; Brunnermeier, Doshi, and James 2018.

58. On standards, see Ding 2021; Ryugen and Akiyama 2020. On great-power competition and standards-setting, see Brunnermeier, Doshi, and James 2018.

59. Arjun Gargeyas, "China's '2035 Standards' Project Could Result in a Technological Cold Warm," *The Diplomat*, September 18, 2021.

60. "China in International Standards Setting," US-China Business Council, February 2020, https://www.uschina.org/sites/default/files/china_in_international_standards_setting .pdf.

61. Gargeyas, "China's '2035 Standards' Project."

62. Liu et al. 2017, 658.

63. Cheung 2022, 25.

64. Doshi 2020, 4.

65. National Science Board 2024. Data from 2021.

66. Kirby 2022, 244.

67. Kirby 2022, 249.

68. World Bank Group 2022.

69. World Bank Group2023. The World Bank defines this statistic as reflecting total enrollment in tertiary education as a percentage of the population in the 5-year age group immediately following upper secondary education.

70. "Record Number of Students to Graduate College in China," Reuters, November 14, 2024. https://www.reuters.com/world/china/record-number-students-graduate-college -china-2025-2024-11-14/.

71. Eberstadt and Abramsky 2022.

72. Zwetsloot et al. 2021, 1.

73. Atkinson and Foote 2019, 9.

74. Cheung 2022; A. B. Kennedy 2019.

75. Kennedy 2018, 19.

76. Fedasiuk, Martinez, and Puglisi 2022, 10.

77. Lind and Mastanduno 2025; Robertson 2023; Zhu et al. 2023.

78. Caroline Wagner, "China's Universities Just Grabbed 6 of the top 10 spots in one worldwide science ranking—without changing a thing," *The Conversation*, April 2, 2024.

79. Wagner, "China's Universities."

80. Times Higher Education 2022.

81. Cited in Kerr and Robert-Nicoud 2020.

82. An Xiao Mina and Jan Chipchase, "Inside Shenzen's Race to Outdo Silicon Valley," *Technology Review*, December 18, 2018.

83. Weinstein et al. 2022.

84. See, for example, Josh Horwitz, "Alibaba Is Plowing $15 Billion into R&D with Seven New Research Labs Worldwide," *Quartz*, October 11, 2017.

85. "China Is the West's Corporate R&D Lab. Can it Remain So?" *The Economist*, July 18, 2024.

86. On the future of Sino-US scientific cooperation given souring political relations, see Natashi Gilbert and Smriti Mallapaty, "US and China Inch Towards Renewing Science- Cooperation Pact—Despite Tensions," *Nature*, September 10, 2024.

87. Elizabeth Gamillo, "China's Artificial Sun Just Broke a Record for Longest Sustained Nuclear Fusion," *Smithsonian Magazine*, January 10, 2022; Darren Orf, "This 'Artificial Sun' Just Smashed Its Own Nuclear Fusion Record," *Popular Mechanics*, January 24, 2025, https://www .popularmechanics.com/science/green-tech/a63512763/nuclear-fusion-east-china/?utm _source=701687&utm_medium=email

88. Woo, Stu Woo, and Clarence Leong, "China Launches Moon Mission in Base Race with US," *Wall Street Journal*, May 3, 2024; Eduardo Baptista, "China Launches Historic Mission to Retrieve Samples from Far Side of the Moon," *Reuters*, May 3, 2024; Smitri Mallapaty, "China's

First Moon Rocks Ignite Research Bonanza," *Nature,* March 15, 2022. Quote from Lyric Li and Christian Davenport, "China Launches World-First Mission to Retrieve Samples from Far Side of Moon," *Washington Post,* May 2, 2024.

89. Li and Davenport, "China Launches World-First Mission."
90. Richard Waters, "US Rushes to Catch Up to China in Supercomputer Race," *Financial Times,* May 17, 2022.
91. Giles 2018.
92. Keith Bradsher, "How China Came to Dominate the World in Solar Energy," *New York Times,* March 7, 2024.
93. Atkinson 2024; Atkinson and Foote 2019, 8; Daisuke Wakabayashi and Claire Fu, "For China's Auto Market, Electric Isn't the Future. It's the Present," *New York Times,* September 27, 2022. On batteries, see Amy Hawkins, "CATL, the Little-Known Chinese Battery Maker That Has the US Worried," *Guardian,* March 18, 2024; Bradsher, "How China Built Tech Prowess." ·
94. Zeyu 2024.
95. Keith Bradsher, "China's Electric Cars Keep Improving, a Worry for Rivals Elsewhere," *New York Times,* May 1, 2024.
96. Diego Mendoza, "China's Lead on EV Battery Innovation Has Not Slipped an Inch," *Semafor,* May 16, 2024.
97. Azusa Kawakami and Yusuke Hinata, "EV Powerhouse China Leads World in Auto Exports," *Nikkei Asia Weekly,* August 6, 2023.
98. Dychtwald 2021.
99. De Smet, Steele, and Zhang, 2021.
100. Chang Che, "All the Drone Companies in China—a Guide to the 22 Top Players in the Chinese UAV Industry," supchina.com, June 18, 2021, https://supchina.com/2021/06/18/all-the-drone-companies-in-china-a-guide-to-the-22-top-players-in-the-chinese-uav-industry/.
101. Zhang 2024; Chorzempa 2022; Dychtwald 2021.
102. Dollar and Huang 2022, 3; Chorzempa 2018; Chorzempa 2022.
103. Keith Bradsher and Joy Dong, "Xi Jinping Is Asserting Tighter Control of Finance in China," *New York Times,* December 5, 2023; on the CCP's reversal of the crackdown, see James Palmer, "Chinese Tech Regulators Back Off," *Foreign Policy,* October 29, 2024, https://foreignpolicy.com/2024/10/29/china-tech-regulation-crackdown-pullback-private-tutoring/; Zhang 2024
104. WIPO 2023.
105. For a discussion of Chinese domestic patenting, see Vagle and Brooks 2025; Ang et al. 2023.
106. Brainard and Normile 2022.
107. The number reflects "a country's share of the top 1% most-cited S&E publications, divided by the country's share of all S&E publications. An index greater than 1.00 means that a country contributed a larger share of highly cited publications; an index less than 1.00 means a smaller share" (National Science Board 2023).
108. Benjamin Plackett, "Nature Index 2024 Research Leaders: India Follows in China's Footsteps as Top Ten Changes Again," *Nature,* June 18, 2024.
109. Bec Crew, "Nature Index 2024 Research Leaders: Chinese Institutions Dominate the Top Spots," *Nature,* June 18, 2024.
110. "The Top 10 Countries for Scientific Research in 2018," *Nature,* July 1, 2019.
111. Luong 2024; Li et al. 2021.
112. Ding 2023, 2024.
113. Paul Mozur and Cade Metz, "In One Key A.I. Metric, China Pulls Ahead of the U.S.: Talent," *New York Times,* March 22, 2024.
114. S. Kennedy 2022. Criticizing the metric of high-value exports are Brooks and Vagle 2025; Beckley 2016.
115. Taylor 2016, 376.
116. "Secretary of State Antony J. Blinken's remarks at the National Security Commission on Artificial Intelligence's (NSCAI) Global Emerging Technology Summit," US Department of State, July 13, 2021.

117. See, for example, Manyika and Spence 2023; Allison et al. 2021; Schwab 2017; Maynard 2015. On general purpose technologies, see Ding 2024.
118. Remarks by the Vice President at the Artificial Intelligence Summit in Paris, France, February 11, 2025, The American Presidency Project, https://www.presidency.ucsb.edu /documents/remarks-the-vice-president-the-artificial-intelligence-action-summit-paris-france.
119. "Global AI Power Rankings: Stanford HAI Tool Ranks 36 Countries in AI," Human-Centered Artificial Intelligence, Stanford University, November 21, 2024, https://hai.stanford .edu/news/global-ai-power-rankings-stanford-hai-tool-ranks-36-countries-in-ai
120. "Global AI Power Rankings"; also see Castro, McLaughlin, and Chivot 2019.
121. Castro, McLaughlin, and Chivot 2019, 2.
122. "China's AI Boom is Reaching Astonishing Proportions," *Economist*, March 11, 2025, https://www.economist.com/business/2025/03/11/chinas-ai-boom-is-reaching-astonishing -proportions; Li et. al 2021.
123. Kyle Wiggers, "Manus Probably Isn't China's Second 'DeepSeek Moment'," *TechCrunch*, March 9, 2025, https://techcrunch.com/2025/03/09/manus-probably-isnt -chinas-second-deepseek-moment/
124. Hannah Beale, "AI Revs Up," *The Wire China*, January 21, 2021, https://www .thewirechina.com/2021/01/03/ai-revs-up/.
125. Emma Farge, "China Leading Generative AI Patents Race, UN Report Says." *Reuters*, July 3, 2024. https://www.reuters.com/technology/artificial-intelligence/china-leading -generative-ai-patents-race-un-report-says-2024-07-03/.
126. Brummer and Lind 2022. Data include AI-related publications in Web of Science database, which received top 1 percent of citations.
127. "The State of Global AI Research," *Emerging Technology Observatory*, Center for Security and Emerging Technologies, Georgetown University, May 5, 2024, https://eto.tech/blog/state -of-global-ai-research/.
128. Evan Osnos, "China's Age of Malaise," *New Yorker*, October 23, 2023; Shirk 2022a; Beckley and Brands 2021.
129. Parts of this discussion were previously published in Lind 2024.
130. Mastro 2024; Biddle and Oelrich 2016; Montgomery 2014; Heginbotham et al. 2015; Anderson and Press 2023.
131. Quoted in Lendon 2018.
132. Mastro 2024, 77.
133. Costello and McReynolds 2018; Pollpeter, Chase, and Heginbotham 2017.
134. On China's previous weakness in its nuclear capabilities, see Lieber and Press 2017, 2006. On China's transformation of its arsenal, see Office of the Secretary of Defense 2022; Stockholm International Peace Research Institute 2024.
135. T. Wright 2021.
136. Quoted in Jeff Seldin, "China Establishing 'Commanding Lead' with Key Military Technologies," *VOA.com*, June 5, 2023.
137. Morgan et al. 2020.
138. Kania 2020.
139. Morgan et. al. 2020, 20.
140. On these programs, see Morgan, et al. 2020; Devin Coldeway, "Carnegie Mellon's Mayhem AI Takes Home $2 Million from DARPA's Cyber Grand Challenge," *TechCrunch*, August 5, 2016.
141. Quoted in Kania 2021a, 520.
142. Kania 2021a, 525.
143. Kania 2020.
144. Morgan et al. 2020, 62.
145. Morgan et al. 2020, 61–62.
146. Ben Noon and Chris Bassler, "Schrodinger's Military? Challenges for China's Military Modernization Ambitions," *War on the Rocks* (blog), October 14, 2021.
147. Buchholz et al. 2020, 9.
148. Buchholz et al. 2020.

149. Kania and Costello 2018.

150. Karen Kwon, "China Reaches New Milestone in Space-Based Quantum Communications," *Scientific American*, June 25, 2020; Giles 2019.

151. US Department of Defense 2021.

152. Aadil Brar, "China and Russia's Unhackable Quantum Satellite Link," *Newsweek*, January 24, 2024.

153. Toner et al. 2023; "China's Tech Crackdown Starts to Ease," *The Economist*, January 19, 2023.

154. Ding 2023.

155. Skinner and Staiger 2007.

156. Wen et al. 2023; Di Wang, Zhou, and Wang 2021.

157. Friedberg 2022; Kennedy 2018; Bader 2018; Economy 2004.

158. Brands and Gaddis 2021; Mearsheimer 2021; Campbell and Ratner 2018.

159. Toner et al. 2023. Allen 2022. The Trump administration subsequently expanded export controls. See Edward Wong, "Trump Makes a New Push to 'Decouple' U.S. From China," *New York Times*, May 30, 2025, https://www.nytimes.com/2025/05/29/us/politics/trump-china-visas-tariffs.html.

160. Brummer and Lind 2022.

161. Brummer and Lind 2022.

162. "US to Indefinitely Extend China Waiver for South Korean Chipmakers, Yonhap Reports," *Reuters*, September 27, 2023; J. Liu and Young 2023. On problems of managing supplier coalitions, see Lind and Mastanduno 2025.

163. Karen Hao, "Huawei Is Giving $300 Million a Year to Universities with No Strings Attached," *MIT Technology Review*, July 3, 2019.

164. Lind and Mastanduno 2025. See also Ting-Fang Cheng, "Huawei Building Vast Chip Equipment R&D Center in Shanghai," *Nikkei Asia*, April 11, 2024.

165. Lind and Mastanduno 2025; Davies 2024; "America's Assassination Attempt on Huawei Is Backfiring: The Company Is Growing Stronger and Less Vulnerable," *The Economist*, June 13, 2024; "Why America's Controls on Sales of AI Tech to China Are so Leaky," *The Economist*, January 21, 2024; Ting-Fang Cheng "How China's Tech Ambitions Slip Through the U.S. Export Control Net," *Nikkei Asia*, October 20, 2023; Ana Swanson, "Takeaways from Our Investigation Into Banned A.I. Chips in China," *New York Times*, August 4, 2024; Raffaele Huang, "China's AI Engineers Are Secretly Accessing Banned Nvidia Chips," *Wall Street Journal*, August 26, 2024.

166. Stephen Nellis, Josh Ye, and Jane Lee, "China's AI Industry Barely Slowed by US Chip Export Rules," *Reuters*, May 3, 2023.

167. Saritha Rai and Julie Zhu, "China's Manus Challenges US Tech Firms in Race to Build AI Agents," Bloomberg.com, March 10, 2025, https://www.bloomberg.com/news/articles/2025-03-10/china-s-manus-challenges-us-tech-firms-in-race-to-build-ai-agents; Gemma Conroy and Smriti Mallapaty, "How China Created AI Model DeepSeek and Shocked the World," *Nature*, January 30, 2025, https://www.nature.com/articles/d41586-025-00259-0; Bloomberg 2023; Hawkins 2023.

168. Kate O'Keeffe, "U.S. Approves Nearly All Tech Exports to China, Data Shows." *Wall Street Journal*, August 16, 2022. Criticizing the Biden administration for inadequate sanctions are Kroenig and Negrea 2024.

169. Gewirtz 2022, 132.

170. World Bank Group 2020; Kennedy 2019; Segal 2011; Liu et al. 2011.

171. Rozelle and Hell 2020.

172. After initially underestimating rising powers, observers often later veer in the opposite direction toward "ten foot tall syndrome" (a problem noted by James Schlesinger, quoted in Hass 2021). For predictions that Japan would overtake the United States, see Kahn 1970; Fingleton 1995.

173. Ben-Atar 2008, 18.

174. Memorandum of Discussion at the 214th Meeting of the National Security Council, Denver, September 12, 1954, *Foreign Relations of the United States, 1952–1954, China and Japan*, Volume XIV, Part II, Document 801, https://history.state.gov/historicaldocuments/frus1952-54v14p2/d801.

175. Breznitz and Murphree 2011, 4.

176. Dan Wang 2023.
177. Woertzel et al. 2015, 36.
178. Woertzel et al. 2015, 42.
179. Dychtwald 2021, 56.
180. Weiss 2014, 1–3.
181. Wade 2017.
182. Chandler 1977; Reich 1983; Hayek 1944.
183. Vallas, Kleinman, and Biscotti 2011; Wade 2017.
184. Weiss 2014, 4.
185. Miller 2022a; Breznitz 2007.
186. Bradsher, "How China Built Tech Prowess"; Groenewegen-Lau 2024; Weinstein et al. 2022; Cheung 2022.
187. Breznitz 2007.
188. "China Has Become a Scientific Superpower," *The Economist*, June 12, 2024; A. B. Kennedy 2019.
189. Acemoglu and Robinson 2012; Kroenig 2020; Brands 2018; Beckley 2021.

3. Smart Authoritarianism

1. Barr 2003.
2. Barr 2003, 300.
3. Morgenbesser 2020; Kerkvliet 2014; Kendall-Taylor and Frantz 2014b; J. Wright 2008; Policzer 2009; Geddes 1999; Krastev 2011; Guriev and Treisman 2022; Levitsky and Way 2002.
4. Aristotle 2013, 166.
5. Rajah 2012, 5.
6. Janos 2000.
7. Huntington 1968, 177-191.
8. Kroeber 2020, 294–95.
9. Ikenberry 2011; Lipset 1959.
10. Kroenig 2020, 21–22.
11. A. H. Zhang 2024 uses this expression in her discussion of Xi Jinping's crackdown in the technology sector; see Chapter 6.
12. This may no longer be true in the Fourth Industrial Revolution; see Beraja et al. 2023.
13. Woertzel et al. 2015; Bhide 2008; Pavitt 1984.
14. For prominent exceptions see Weeks 2014; Talmadge 2015; Reiter 2020.
15. On a "democratic difference" see, for example, Bueno de Mesquita, et. al 1999; Fearon 1994; Doyle 1986.
16. On the distinction between institutional constraints versus policies (or "practices"), see Glasius 2018b; Pepinsky 2020; Przeworski and Limongi 1993; Glaser et al. 2004.
17. Gwartney, Lawson, and Holcombe 1999, 658.
18. Przeworski and Limongi 1993, 52.
19. Glaser et al. 2004, 277.
20. Glaeser et al. 2004, 298.
21. J. Wright 2008, 342; Gandhi 2008a; Weede 1996.
22. Weede 1996.
23. Sinkkonen 2021; Geddes, Wright, and Frantz 2018; Gandhi 2008b; Geddes 1999; Morgenbesser 2020; Hadenius and Toerell 2007.
24. Haggard 1990, 263.
25. Dobson 2013, 4–5.
26. Tip of the hat to Krastev 2011; also on these categories, see Levitsky and Way 2002; Diamond 2002; Kendall-Taylor and Frantz 2014b; Morgenbesser 2020.
27. Cooley and Snyder 2015; Marinov and Goemans 2014.
28. On authoritarian participation in international justice efforts, see Subotić 2009. On countries seeking ascension to the European Union, see Gray 2009. On women's rights in

authoritarian regimes, see Bjarnegård and Zetterberg 2021, 2022; Bush 2011; Donno, Fox, and Kaasik 2022. On the effects of election monitoring, see Asunka et al 2019. On the diminished rewards to military coups d'etat, and the ensuing reduction in coups, see Kendall-Taylor and Frantz 2014a; Marinov and Goemans 2014.

29. Ryan 2022, 152.

30. See, for example, Daemmrich 2017; Schwab 2017.

31. Kuhn 1962; Hanchen Wang et al. 2023. I am grateful to Matthew Brummer for discussion of these points.

32. Hanchen Wang et al. 2023.

33. Olar 2019; Tansey 2016; Hall and Ambrosio 2017; Heydemann and Leenders 2011.

34. Frantz 2018, 66.

35. Frantz 2018, 67; Kendall-Taylor and Frantz 2014b; Gandhi 2008b; Gandhi and Przeworski 2007.

36. Frantz 2018, 67, 120.

37. Gandhi 2008a, xviii.

38. Meng 2020; Boix and Svolik 2013; Svolik 2012; Gandhi 2008b.

39. Gandhi 2008b, 78.

40. Frantz 2019.

41. Terminology from Levitsky and Way 2002.

42. Guriev and Treisman 2019, 2022.

43. Quoted in Guriev and Triesman 2019.

44. Derby 2009, 2–3; Guriev and Treisman 2022.

45. Quoted in Patrick Kingsley, "On the Surface, Hungary is a Democracy. But What Lies Underneath?" *New York Times*, December 25, 2018.

46. Guriev and Triesman 2019.

47. Guriev and Treisman 2022; Olar 2019; Frantz 2018; Kendall-Taylor and Frantz 2014; Policzer 2009.

48. Frantz 2018, 107.

49. Kerkvliet 2014, 2019; Guriev and Treisman 2022.

50. Ibrahim 2021.

51. Levitsky and Way 2002, 53; Kim and Gandhi 2010.

52. Dobson 2013, 51–52.

53. Kerkvliet 2014, 102.

54. Kerkvliet 2014, 102.

55. Guriev and Treisman 2022, 48–49.

56. Guriev and Treisman 2022, 50.

57. Kerkvliet 2014, 115; 2019.

58. Przeworski 2004.

59. Wedeman 1997; Glaeser et al. 2004, 276.

60. On horizontal versus vertical accountability, see Diamond and Morlino 2004.

61. Quotes from Rajah 2012, 8; Tamanaha 2004, 5.

62. Moustafa 2007.

63. Sievert 2018, 774.

64. Yuhua Wang 2015, 4.

65. Silverstein 2008, 74, 83.

66. Chungshik Moon 2019, 1258.

67. Mkandawire 2015; Lienert and Modi 1997; Van de Walle 2001.

68. Rogoff 1985; Alesina and Summers 1993; Acemoglu, et al. 2008b.

69. Arias, Hollyer, and Rosendorff 2018, 907; see also Gehlbach and Keefer 2011, 2012; Jensen, Malesky, and Weymouth 2014; Fang 2011. Asserting the inability of autocracies to make binding agreements are Acemoglu et al 2008.

70. On illiberal autocratic trade policy, see Bhagwati 1978; Shatz and Tarr 2000; Todaro 1977.

71. Nelson and Pack 1999.

72. Rock 2017, 46.

73. Mkandawire 2015; on Latin America, see Schneider 1999; on Nigeria, see Ugoani 2016.

74. Kang 2002; Wedeman 1997; Laothamatas 1994. On types of corruption that are more damaging versus less damaging to economic growth, see Ang 2020.
75. Evans and Rauch 1999; Kang 2002; Haggard 1990.
76. Robert F. Worth, "Mohammed Bin Zayed's Dark Vision of the Middle East's Future," *New York Times*, January 9 2020.
77. Robbie Gramer, "Biden Taps Billionaire Campaign Donors for Ambassador Posts," *Foreign Policy*, December 20, 2021; Fedderke and Jett 2017.
78. Katz 1998, 154.
79. Green 2011, 434; Gil Mulligan, and Sala-i-Martin 2004.
80. Lee 1997; McGuire 2010, 210.
81. Worth, "Mohammed Bin Zayed's Dark Vision."
82. Arguing that authoritarian regimes take care to restrict "coordination goods" that enable people to mobilize against the regime are Bueno de Mesquita and Downs 2005.
83. Koh Tai Ann (1998), quoted in T. Lee 2002, 102.
84. Gavin 2024 (Georgia); Krekó 2017 (Hungary); "Laws of Attrition"; "Kyrgyzstan"; "Nicaragua Shuts Down 50 Non-Profits in New Crackdown," *BBC.com*, May 5, 2022.
85. Frantz 2018, 67.
86. Dobson 2013, 27–28.
87. Egorov, Guriev, and Sonin 2009.
88. Another tactic that squares this circle, as I discuss in Chapter 4, is a dual media. In China, the public media emphasizes positive news and censors discussion of national problems, whereas an internal media—which circulates on a highly curtailed basis to high-level government officials—includes frank investigative reporting and discussion of national problems and government failings. See Dimitrov 2017.
89. Dobson 2013, 111.
90. Guriev and Treisman 2022, 86–87.
91. On bribery, see McMillan and Zoido 2004; on Ivcher's case, see "Peru Strips Controversial TV Mogul."
92. On the Hungarian media, see Kingsley, "On the Surface, Hungary is a Democracy"; Zack Beauchamp, "It Happened There: How Democracy Died in Hungary," *Vox.com*, September 13, 2018.
93. Ira Glass, "Do Not Go Gentle," *This American Life* no. 767, National Public Radio, April 8, 2022, https://www.thisamericanlife.org/767/transcript.
94. Roberts 2018.
95. Roberts 2018, 6.
96. Lu and Pan 2021.
97. Yang, forthcoming.
98. Chandra and Rudra 2015, 990.
99. Policzer 2009; Teets 2014.
100. Lorentzen 2013, 129.
101. Weede 1996.
102. Hankla and Kuthy 2013, 212.
103. Beckley, Horiuchi, and Miller 2018.
104. United Nations 2022.
105. Abuza 1996, 620.
106. Krastev 2011, 8.
107. Guriev and Triesman 2022, x.
108. Glasius 2018a.
109. Dukalskis 2021.
110. Tsourapas 2019.
111. Michaelsen 2018.
112. Deibert et al. 2011.
113. Amy Mackinnon and Mary Yang, "The Booming Export of Authoritarianism," *Foreign Policy*, June 2, 2022.
114. Al-Rawi 2021.

115. Antonio Panzeri, "The E.U. Must Stand for Human Rights in Saudi Arabia—and Justice for Jamal Khashoggi," *Washington Post*, March 12, 2019.

116. Tornell 1995; Gourevitch 1986.

117. Ravina 2017.

118. Schleunes 1979, 324.

119. Lincoln 1990.

120. Waterbury 1992, 183.

121. Herbst 2021.

122. Overholt 2018, 9–10.

123. Quoted in Chung-in Moon 1994; Haggard, Pinkston, and Seo 1999.

124. Olson 1965.

125. Haggard 2000, 27.

126. Haggard, Pinkston, and Seo 1999, 212.

127. Schamis 1999, 238; Sun 1999.

128. Chung-in Moon 1994, 150.

129. Kingstone 2010.

130. Guriev and Papaioannou 2022, 758; also see Dornbusch and Edwards 1991.

131. For an argument that China's government suffers from short-termism, see Ji 2023.

132. Moyo 2018.

133. Galston 2015.

134. Lindblom 1977, 346.

135. Lindblom 1977, 196.

136. For a survey of "authoritarian advantage" arguments, see Drezner 2022; Kroenig 2020. A version of such arguments is seen in discussion of weak property rights in China, which enable the government to acquire massive amounts of data that will help its firms develop artificial intelligence technology. See Li, Tong, and Xiao 2021; Zeng 2020.

137. Bhagwati 1966, 203–204.

138. Khanin 2003, 1199.

139. Financial repression entails policies that reduce the people's income and their choices about how to save and spend that income. Policies, for example, include banning private banks and foreign financial firms (giving people no choice but to save in banks controlled by the government), imposing strict capital controls (so people can't invest overseas), and manipulating currency and deposit rates. For an excellent discussion of such policies in China, see Pettis 2011.

140. Wade 1990; C. Johnson 1982; Haggard 1990. Also see O'Donnell and Schmitter 1986; Kohli 2004.

141. Huntington 1968; Wade 1990.

142. Schmidt and Bajraktari 2022.

143. On industrial policy, see Juhász, Lane, and Rodrik 2024. On countering China with a US industrial policy, see Patricia Cohen, Keith Bradsher, and Jim Tankersly, "How China Pulled So Far Ahead on Industrial Policy," *New York Times*, May 27, 2024; Siripurapu and Berman 2023; Miller 2022b.

144. Kastner and Wohlforth 2021.

145. Pickard 2019; Jerit and Zhao 2020.

146. Ma's career was reinvigorated after Alibaba's focus on artificial intelligence—a priority sector for the CCP. See Zijing Wu and Eleanor Olcott, "How Jack Ma's Pivot to AI Rehabilitated Alibaba." *Financial Times*, March 16, 2025. https://www.ft.com/content/df2bccee-1730-402f-bb92-9d743018324f.

147. For an argument about media freedoms in resource-poor countries, see Egorov, Guriev, and Sonin 2009.

148. Ermolaev 2019; Aven, Nazarov, and Lazaryan 2016.

149. Heim and Salimov 2020.

150. Doner, Ritchie, and Slater 2005, 328.

151. For a similar argument applied to Rwanda, see Mann and Berry 2016; Green 2011. For arguments about external threats and development, see Kang 2002.

152. Gallagher and Hanson 2013.

153. Taylor 2016. On external threats and technology absorption see Milner and Solstad 2021.

154. For a similar argument about domestic politics and innovation, see Doner, Hicken, and Ritchie 2009.

155. Henry Porter and Annabel Davidson, "Colonel Qaddafi—A Life in Fashion," *Vanity Fair*, August 12, 2009. https://www.vanityfair.com/news/photos/2009/08/qaddafi-slidesho w200908?srsltid=AfmBOoqyWMChW3BX6TjPIc8Fla65-axOYWGzs8STn4vSs536A0fdAgDq.

156. Hanson and Sokhey 2021; Lorch and Bunk 2017; on the taming of the middle class, see Rosenfeld 2017, 2020.

4. China Gets Smart

1. Zeng 1887. Zeng is also known as Marquis Tseng.

2. Wakeman 1997.

3. M. C. Wright 1957, 177; Schmid and Huang 2017, 575.

4. Brodie and Brodie 1973; W. A. Murray 2005; Boot 2006.

5. Bloom and Williamson 1998; Bloom, Canning, and Sevilla 2003; Stevenson and Wang 2023; Fang 2016; Goldstone et al. 2012; Madsen 2012.

6. Terminology from Rostow 1960.

7. Babiarz et al. 2015.

8. Babiarz et al. 2015.

9. Babiarz et al. 2015.

10. Sidel 1972; Zhang and Unschuld 2008.

11. K. A. Johnson 1985, 9.

12. Kellee Tsai 1996, 499.

13. "China's New Marriage Law," 369.

14. Niida 1964.

15. Wilson 2012; Mackie 1996. Foot binding had been outlawed in 1912 but the custom remained widely practiced until after 1949.

16. Lavely et al. 1990, 70, 72. The authors note that the famine years halted progress in female rural education; "The heyday of progress in primary education was the 1950s, when growth stood at 3 per cent or above for seven years."

17. Lavely et. al. 1990, 68.

18. Feng 2011, 174.

19. Rodriguez 2023.

20. Wang and Mason 2007.

21. Ma 2013.

22. Whyte, Feng, and Cai 2015.

23. Vermeer 2006.

24. K. Zhou 1996, 182.

25. Whyte, Feng, and Cai 2015; Feng 2011, 175.

26. Whyte, Feng, and Cai 2015.

27. Whyte, Feng, and Cai 2015.

28. Fong 2016.

29. Khan and Hu 1997. K. Zhou (1996, 7) notes that rural industry grew upward at 20 percent in the 1970s.

30. Data in 2015 constant $USD, from World Bank Group, 2024.

31. Wallace 2022.

32. Naughton 2008.

33. Zhu 2012, 103.

34. K. Zhou 1996, 5.

35. Gewirtz 2022, 17.

36. K. Zhou 1996, xxix.

37. Shirk 1993, 23; Yang 1996.
38. Guo 2003, 11.
39. Quoted in Guo 2003, 11. On the crisis of the Cultural Revolution, see Gewirtz 2022, 18; Feng Wang 2024.
40. King, Pan, and Roberts 2013; Lorentzen 2014; Dimitrov 2017.
41. Dimitrov 2017.
42. Shirk 1993, 12.
43. Woo 1994, 304.
44. Naughton 2008, 93.
45. Naughton 2008, 113.
46. Naughton 2008, 98.
47. Shirk 1993, 35.
48. Gewirtz 2022, 18.
49. Quoted in Shirk 1993, 35.
50. Schell and Delury 2014, 14.
51. Taiwan growth statistics from P. Tsai 1999.
52. Data from 1980 from International Monetary Fund 2025.
53. Lieberthal 1978.
54. Gewirtz 2022; Vogel 2013.
55. Gewirtz 2022, 205–10.
56. Shambaugh 2016, 98–99.
57. K. Zhou 2009.
58. Shirk 1993, 34.
59. Mitter 2003, 120.
60. Shirk 1993, 129.
61. Quoted in Shirk 1993, 130.
62. Shirk 1993, 130.
63. Naughton 1995; Lau, Qian, and Roland 2008.
64. Judy Heflin, "The Single Greatest Educational Effort in Human History," *Language*, https://www.languagemagazine.com/the-single-greatest-educational-effort-in-human-history/
65. Holm 1991.
66. Chen 1980.
67. Peterson 1994, 101.
68. Peterson 1994, 113. Peterson notes that in its 1956 literacy campaign announcement, the Party defined *literacy* in terms of the ability to recognize 1,500 characters, read books and magazines, understand an account book, and use an abacus for simple calculations.
69. Cleverley 1991.
70. Heflin, "The Single Greatest Educational Effort."
71. Fairbank 1971, 33.
72. Lauwerys and Huq 2024.
73. Lavely et al. 1990; Saywell 1980.
74. Quoted in Evan Osnos, "The Cost of the Cultural Revolution, Fifty Years Later," *New Yorker*, May 6, 2016; also see Liqing Tao, Margaret Berci, and Wayne He, "Historical Background: Expansion of Public Education," *New York Times*, June 18, 1916.
75. Hannum et al 2008.
76. Rosen 1984.
77. Mok 2007.
78. Tao, Berci, and He, "Historical Background."
79. Kirby 2022.
80. Quoted in S, Han and Xu 2019; also see Yang and Currie 2006.
81. Han and Xu 2019, 935.
82. J. Wright 2008; Gandhi 2008a.
83. Quoted in A. Cook 2016.
84. Law 2002.

85. Quoted in A. Cook 2016.
86. Shirk 1993, 8.
87. Naughton 2008, 112–113.
88. Landry 2008.
89. Guo 2006, 160.
90. Quoted in Suzuki 2018.
91. Nathan 2003, 7.
92. Vogel 2013.
93. Bian 1994.
94. On Chinese civil service reform, see Burns 1989, 2003.
95. Goldman and MacFarquhar 1999, 14.
96. Goldman and MacFarquhar 1999, 13.
97. Goldman and MacFarquhar 1999.
98. Goldman and MacFarquhar 1999, 12.
99. Naughton 2008, 101–102.
100. Zhou and Xin 2020, 96; also see Shambaugh 2022.
101. The same was true in the Soviet Union, which developed world-class technology in the early post–World War II years despite a lack of (Western) intellectual property rights. See Soltysinski 1969.
102. Law 2002; Yuhua Wang 2015, 51.
103. Li 2004, 106.
104. Kaufman 1998, 76.
105. Brandt and Rawski 2008, 43.
106. Brandt and Rawski 2008, 19.
107. Awokuse and Yin 2010.
108. Yuhua Wang 2015, 32; Hsiao and Hsaio 2004.
109. Awokuse and Yin 2010.
110. Che and Qian 1998.
111. Hou 2019, 2.
112. See, for example, P. Ho 2001.
113. Heston and Sicular 2008, 50.
114. Branstetter and Lardy 2008, 652.
115. D. Li 1996; Walder and Oi 1999; Francis 2001; Ang 2016.
116. Montinola, Qian, and Weingast 1995, 58.
117. Clarke, Murrell, and Whiting 2008, 402.
118. Du, Lu, and Tao 2008; Awokuse and Yin 2010.
119. Clarke, Murrell, and Whiting 2008, 402.
120. Che and Qian 1998, 491.
121. Hou 2019, 12.
122. Ang 2016.
123. Mkandawire 2015; Green 2011; Wedeman 1997.
124. Ang 2020; Wedeman 1997; Wedeman 2012.
125. Kroeber 2020, 208.
126. "How Does Corruption Hinder China's Development?"
127. Ang 2020, 11–12.
128. Ang 2020, 12–13.
129. Bai, Hsieh, and Song 2020.
130. Naughton 2008, 110.
131. Whiting 2001.
132. Naughton 2006, 112.
133. Pei 2006.
134. Li 2013.
135. Ang 2020, 3–5.
136. Emily Feng, "How China's Massive Corruption Crackdown Snares Entrepreneurs Across the Country," *NPR.org*, March 4, 2021.

137. Kroeber 2020, 208–9; Shambaugh 2015. On the economic effectiveness of these reforms see Wedeman 2017.
138. Dikötter 2019, 99–100.
139. Dikötter 2019, 103–4.
140. Dikötter 2019, 110.
141. Yang 2011; Thurston 1987; Walder and Su 2003; Youqin Wang 2001.
142. Walder 2014, 522.
143. Tanner 2000, 111.
144. Ang 2022.
145. Pei 2000, 27.
146. "Chinese Journalist Mao Huibin Arrested After Publishing Articles about Tangshan Assault," Committee to Protect Journalists, August 15, 2022, https://cpj.org/2022/08/chinese-journalist-mao-huibin-arrested-after-publishing-articles-about-tangshan-assault/#:~:text=According%20to%20CPJ's%20most%20recent,third%20year%20in%20a%20row.
147. Wang and Minzner 2015.
148. Guo 2012.
149. Wang and Minzner 2015.
150. O'Brien and Deng 2017, 181.
151. Lorentzen 2013, 148.
152. Bernstein and Lu 2003.
153. Li 2019.
154. Li 2019, 310.
155. Lorentzen 2013, 130.
156. Lorentzen 2013, 129.
157. Li 2019.
158. Greitens et al. 2020, 9, 45.
159. Mattis 2011; also see Schwarck 2018.
160. O'Brien and Deng 2017, 180, 182.
161. Truex 2019, 1033.
162. Chan and Qiu 2003.
163. Zhou 1996, xxiii.
164. Gilboy and Read 2008, 145.
165. Tong 1994, 337.
166. Chen and Dickson 2010; Dickson 2007.
167. Gilboy and Read 2008.
168. Saich 2000, 124; Gallagher 2004; Teets 2014; Frolic 1997.
169. Shaun Shieh, "Remaking China's Civil Society in the Xi Jinping Era," ChinaFile, August 2, 2018; also see K. Zhou 2009.
170. Shirk 2010, 7.
171. Nathan 2003, 12.
172. Shirk 2010, 9.
173. Tong 1994, 336.
174. "China Media Guide," BBC.com, August 22, 2023.
175. Shirk 2010, 5.
176. Kurlantzick 2003, 52.
177. Hildebrandt 2013, 1.
178. Saich 2000.
179. Kurlantzick 2003, 52.
180. Matthew Pottinger, quoted in Beina Xu 2014.
181. Tommy Walker, "Most Countries in Asia See Decline in Press Freedom," VOA.com, May 3, 2024.
182. Han and Shao 2022; Lorentzen 2014.
183. Roberts 2018, 5–6.
184. Lu and Pan 2021.

185. Wang and Mark 2015.
186. Nixon 1967, 121.
187. Spence 1999a, 153.
188. Spence 1999a, 136.
189. Spence 1999b, 508.
190. Hu and Jefferson 2008, 292.
191. Mao 1977, 103.
192. Eckstein 1977, 238.
193. Wu 1981.
194. Wu 1981, 471.
195. Quoted in Wu 1981, 478.
196. Heston and Sicular 2008, 47.
197. Wu 1981, 473.
198. Data in this paragraph from Richter 1989, 25–26.
199. Richter 1989, 29.
200. Zweig and Rosen 2003.
201. Zou Shuo, "Overseas Study No Longer Only for the Elite," *China Daily*, September 19, 2019, https://www.chinadaily.com.cn/a/201909/19/WS5d82e054a310cf3e3556c563_3.html.
202. Hannum et al. 2008, 234; "Statistics on Chinese Learners Studying Overseas in 2019."
203. K. Zhou 2009, xxiii, xxix.
204. I am grateful to an anonymous reviewer for this point.
205. Kellee Tsai 1996; Zheng 2010.
206. De Bary and Lufrano 1999, 243.
207. Gewirtz 2022.
208. Shirk 2022; Shambaugh 2021; Magnus 2018; Economy 2018; Rudd 2022; Ian Bremmer, "'Maximum Xi Jinping' Will Be Costly for China and Its People," *Nikkei Asia*, January 18, 2023.

5. Crazy Rich Authoritarians

1. Dikötter 2019; Szabłowski 2020.
2. Bueno de Mesquita and Smith 2011; Levitsky and Ziblatt 2018. Also see Mounk 2018b.
3. See, for example, Mike Godwin, "Yes, It's Okay to Compare Trump to Hitler. Don't Let Me Stop You," *Washington Post*, December 20, 2023; Jonathan Blitzer, "A Scholar of Fascism Sees a Lot That's Familiar with Trump," *New Yorker*, November 4, 2016.
4. Material in this section was previously published in Lind 2023.
5. Schleunes 1979, 322. On the rejection of public education in Qing-dynasty China, see Borthwick 1978, 25–26.
6. Acemoglu and Robinson 2012, 226.
7. Acemoglu and Robinson 2012, 225.
8. Wakeman 1997.
9. Brodie and Brodie 1973; Murray 2005; Boot 2006.
10. On technology and the decline of the Ottoman Empire see Lewis 1958; Pamuk and Williamson 2011.
11. Levinger 2000.
12. Schleunes 1979, 318; Lohmann and Mayer 2007.
13. Schleunes 1979, 317.
14. Ding 2021; Mokyr 1993.
15. Mosse 1966.
16. Mosse 1966, 83.
17. Pereira 1983, 68.
18. Mosse 1966.

19. Mosse 1966, 88.
20. For discussion of how education promoted national unity, see Weber 1976; Ravina 2017.
21. Schleunes 1979, 336.
22. K. Y. Lee 2000, 44, 52.
23. Haggard 1990, 102.
24. Vasagar 2022, 8.
25. K. Y. Lee 2000, 5; Vogel 2011, 529; Chee 1971.
26. World Bank Group. 2025.World Development Indicators. "GDP per capita GDP (current $US)," Data.worldbank.org, accessed March 24, 2025.
27. Data from https://www.worldbank.org/en/country/singapore/overview#1.
28. "Singapore" 2021, 2023.
29. "Singapore's Global Innovation Ranking" 2024.
30. "Singapore Ranking in the Global Innovation Index" 2023.
31. On demographic transitions, their causes, and their economic effects, see Goldstone, Kaufmann, and Toft 2012; Bloom, Canning, and Sevilla 2003; Bloom and Williamson 1998.
32. Saw 2012; Pyle 1997; Moi 2015; Goh and Gopinathan 2008.
33. Leong 2008.
34. Kum 2008.
35. Fawcett and Koo 1980, 555.
36. Pyle 1997, 216.
37. All in all, Singapore's government was *too* successful in bringing down the fertility rate; worries shifted from overpopulation to the birth rate being inadequate, and the PAP later shifted to pronatal policies. See Palen 1986.
38. Tan 2012, 71.
39. George 2007, 133.
40. Tey 2008, 618.
41. Rodan 1996, 72–73.
42. "Singapore," Freedom in the World 2021.
43. Wee, Sui-Lee Wee, "Singapore's Riches Grew Under its Leader. So Did Discontent," *New York Times*, May 14, 2024.
44. Quah 1999, 492.
45. Huff 1995, 1433.
46. Vasagar 2022, 12.
47. Zee 2011.
48. Haggard 1990, 105.
49. George 2007, 133.
50. *The Economist* 2008.
51. Alexander 2013.
52. Quoted in Alexander 2013.
53. George 2007, 135.
54. Vasagar 2022, 183.
55. Vasagar 2022, 183–84.
56. Jonathan Lin, "How Much Freedom Does the Press Have in Singapore, Explained." *The Kopi*, May 12, 2021, https://thekopi.co/2021/05/12/press-freedom-singapore/.
57. Lin, "How Much Freedom."
58. U.S. Department of State, "Country Reports on Human Rights Practices for 2016."
59. "Singapore: Social Media Companies Forced to Cooperate with Abusive Fake News Law."
60. Tan 2020, 1073 N. T. Kadir 2020, 1073; also S. Kadir 2004.
61. Vasagar 2022, 188.
62. "'Kill the Chicken to Scare the Monkeys'" 2017, 8.
63. "'Kill the Chicken to Scare the Monkeys'" 2017.
64. World Bank Group 2023b.
65. "Best Universities in Singapore 2024," *Times Higher Education World University Rankings*, https://www.timeshighereducation.com/student/best-universities/best-universities-singapore.

66. Koh and Soon 2012.
67. S. Kadir 2004, 330.
68. S. Kadir 2004, 350
69. K. Y. Lee 2000, 135.
70. K. Y. Lee 2000, 136
71. Goh and Gopinathan 2008, 84.
72. Goh and Gopinathan 2008, 84.
73. Goh and Gopinathan 2008, 84.
74. Quoted in Ooi 2010, 227.
75. Krugman 1994.
76. Program for International Student Assessment Scores, reporting testing of 15-year-olds by country. See https://factsmaps.com/pisa-2018-worldwide-ranking-average-score-of-mathematics-science-reading/.
77. Ooi 2010, 158.
78. Rana and Lee 2015.
79. K. Y. Lee 2000, 135.
80. Vogel 2011, 532; K. Y. Lee 2000.
81. K. Y. Lee 2000, 61. Also see Poh-Kam Wong 1988.
82. Rajah 2012, 3.
83. Silverstein 2008, 74, 83.
84. Wee, "Singapore's Riches Grew."
85. Karishma Vaswani, "Taylor Swift Is Helping Singapore Ditch Its Dull Reputation," *Bloomberg Opinion*, January 7, 2024.
86. Rajaratnam 1972.
87. Ooi 2010, 154.
88. Gooroochurn and Sugiyarto 2005.
89. Wong and Tang 2010, 966.
90. Ooi 2010, 152–53.
91. Hirschman 1968; Haggard 1990.
92. Ooi 2010, 117.
93. Pace 1984.
94. K. Y. Lee 2000.
95. K. Y. Lee 2000.
96. "Raffles Statue." On "Rhodes Must Go," see Amitabh Chaudhuri, "The Real Meaning of Rhodes Must Fall," *Guardian*, May 16, 2016. Making a similar argument about the Raffles statue is Vasagar 2020.
97. E. W. Ho 2006, 388.
98. Vogel 2011, 532.
99. Kaplan 2023, 191.
100. Nazarbayev 2008, 5.
101. Ahn, Dixon, and Chekmareva 2018, 202.
102. Yakavets and Dzhadrina 2014, 30.
103. Ahn, Dixon, and Chekmareva 2018, 227.
104. "The World Bank in Kazakhstan," World Bank, https://www.worldbank.org/en/country/kazakhstan/overview.
105. World Development Indicators 2023, World Bank.
106. GDP per capita expressed in constant 2015 USD. World Development Indicators 2023, World Bank
107. Cornell, Starr, and Barro 2021; "A New Course for the Republic."
108. Chris Rickleton, "Kazakh Opposition Candidates Battle to Compete in Parliamentary Elections," *Radio Free Europe*, March 3, 2023.
109. Quoted in Rickleton "Kazakh Opposition Candidates." Akhmedyarov was a journalist who had been detained and tortured by the government. He survived an assassination attempt in 2012.

110. Aliya Tlegenova and Serik Beysombaev, "Have President Tokayev's Reforms Delivered a 'New Kazakhstan'?" Carnegie Endowment for International Peace, September 18, 2024, https://carnegieendowment.org/russia-eurasia/politika/2024/09/tokayev-kazakhstan -reforms?lang=en.

111. Bakhyzhan Kurmanov, "The Rise of 'Information Autocracies': Kazakhstan and Its Constitutional Referendum," The Loop blog, European Consortium for Political Research, June 7, 2022, https://theloop.ecpr.eu/the-rise-of-information-autocracies-kazakhstan-and-its -constitutional-referendum/; Schiek 2022.

112. Wood 2023, 50.

113. Wood 2023.

114. Wood 2023, 45.

115. Sharivkan et al. 2016.

116. "Kazakh President Signs into Law Controversial Bill on Media," *Radio Free Europe*, June 20, 2024. On media control, see Anceschi 2015.

117. Schiek 2018.

118. Luca Anceschi, "Modernizing Authoritarianism in Uzbekistan," *Open Democracy*, July 9, 2018, https://www.opendemocracy.net/en/odr/modernising-authoritarianism-in-uzbekistan/; Anceschi 2015.

119. Anna Bjerde, "To Withstand Global Shocks, Uzbekistan Needs to Continue Reforms and Build and Inclusive Market Economy," *The Diplomat*, November 1, 2022.

120. Ottaway and Dunne 2007.

121. Peck 2024.

122. Mati and Rehman 2022; Fatma Tanis, "Saudi Arabia Sees Massive Cultural Shift After Crown Prince's Reforms," *NPR.org*, August 7, 2022.

123. Neve 2021; John Hannah, "Washington Is Oblivious to the Importance of Saudi Reforms," *Foreign Policy*, January 16, 2023.

124. Hannah, "Washington Is Oblivious."

125. Arezki 2024.

126. Uniacke 2022.

127. Al-Rasheed 2021, 9.

128. "As Repression Grows in Saudi Arabia, So Too Does Investment in Sports," CBS News, *60 Minutes Overtime*, April 9, 2023.

129. Julian E. Barnes and David E. Sanger, "Saudi Crown Prince Is Held Responsible for Khashoggi Killing in U.S. Report," *New York Times*, February 26, 2021.

130. Arezki 2024.

131. "How Saudi Arabia Is Reforming Education Through Technological Investment."

132. "How Saudi Arabia Is Reforming Education Through Technological Investment."

133. "UN Tourism Applauds Saudi Arabia's Historic Milestone of 100 Million Tourists." UN Tourism, 2024.

134. "After Decades of Empty Talk, Reforms in Gulf States Are Real—But Risky," *The Economist*, February 9, 2023.

135. D. Ottaway 2021.

136. Nathan French, "Saudi Reforms Are Softening Islam's Role, But Critics Warn the Kingdom Will Still Take a Hard Line Against Dissent," *The Conversation*, September 5, 2023.

137. Peck 2024.

138. "Economic Diversification Efforts Paying Off in GCC Region but More Reforms Needed" 2023.

139. "How the Authoritarian Middle East Became the Capital of Silicon Valley," *Washington Post*, May 14, 2024.

140. Maureen Farrell and Rob Copeland, "Saudi Arabia Plans $40 Billion Push into Artificial Intelligence," *New York Times*, March 19, 2024. https://www.nytimes.com/2024/03/19 /business/saudi-arabia-investment-artificial-intelligence.html.

141. Robert D. Kaplan, "Saudia Arabia Channeling Singapore," *National Interest*, August 22, 2023.

142. Stephen Kalin, "Saudi Arabia Wants to Export Its Economic Reforms, Finance Minister Says," *Wall Street Journal*, January 20, 2023.
143. Allen et. al 2025.
144. Worth 2020.
145. Yom 2009, 151–152.
146. Yom 2009, 155.
147. Yom 2009.
148. Beck and Hüser 2015.
149. Wiktorowicz 2000.
150. Nanes 2008, 81.
151. Yom 2009.
152. C. W. Jones 2017, 37.
153. Ottaway and Dunne 2007, 19.
154. Fakir and Werenfels 2020.
155. Economic data in constant $USD from the Federal Reserve Bank of St. Louis, https://fred.stlouisfed.org/series/NYGDPPCAPKDMAR.
156. Fakir 2021.
157. Fakir 2021.
158. Loudiy 2014.
159. Ottaway and Dunne 2007, 10.
160. Matfess 2015.
161. Tapscott 2021; also see Ong 2022.
162. Abrahamsen and Bareebe 2021.
163. Abrahamsen and Bareebe 2021.
164. "Africa's Tiger Economy Is Shot," *The Economist*, February 29, 2024.
165. "The Violence in Ethiopia Imperils an Impressive Growth Record," *The Economist*, November 20, 2021.
166. "The World Bank in Ethiopia," https://www.worldbank.org/en/country/ethiopia/overview.
167. Giulia Paravicini, "In Ethiopia, a Secret Committee Orders Killings and Arrests to Crush Rebels," *Reuters*, February 23, 2024; "A Hidden War Threatens Ethiopia's Transition to Democracy," *Economist*, March 19, 2020.
168. "War Crimes in Tigray May Be Covered Up or Forgotten," *The Economist*, July 9, 2023.
169. Gardner 2022.
170. Meester 2021.
171. "The Historic Heart of Addis Ababa Is Being Demolished," *The Economist*, April 24, 2024.
172. "Africa's Tiger Economy Is Shot," *The Economist*, February 29, 2024.
173. Zack Beauchamp, "It Happened There," *Vox*, September 13, 2018; Yascha Mounk, "Poland's Imperiled Democracy," *Atlantic*, June 8, 2023; Gabriela Baczynska, "Poland, Hungary Turning More Authoritarian, Rights Group Says," *Reuters*, February 15, 2022; Kakachia and Lebanidze 2023.

6. Staying Smart

1. On the shift in Chinese politics and its implications for China's future, see ; Shirk 2022a; Rudd 2022; Pearson, Rithmire, and Tsai 2022; Shambaugh 2021; Magnus 2018; Economy 2018.
2. Lardy 2024, 2019.
3. Bergsten 2022.
4. Quoted in Allison and Blackwill 2013. Italics in original.
5. I previously discussed these points in Lind 2024.
6. Pritchett and Summers 2014; Eichengreen, Park, and Shin 2013.
7. On input-based growth and why it slows, see Young 1995; Krugman 1994.
8. On the middle-income trap, see "The High Kingdom," *The Economist*, February 5, 2022; M. Levy 2022; Foxley and Sossdorf 2011.

9. Kharas and Kohli 2011, 282.

10. World Bank, *World Development Indicators*, "GNI per Capita, Atlas Method (current US$) – China," 2024, https://data.worldbank.org/indicator/NY.GNP.PCAP.CD?locations=CN.

11. World Bank, *World Development Indicators*, accessed March 2025, "GNI per Capita, Atlas Method (current US$)," 2024, https://data.worldbank.org/indicator/NY.GNP.PCAP.CD.

12. Lind 2024.

13. Vagle and Brooks 2025; Lind 2024.

14. Beckley and Brands 2022, 2021; Brands and Beckley 2021; Shambaugh 2016; Shirk 2022a; Economy 2011.

15. Stevenson and Wang 2023; Fang 2016; Goldstone, Kaufmann, and Toft 2012; Madsen 2012.

16. Pretransition families typically had 7–8 children; parents wanted 3–4 children but knew that because of high mortality rates, many of them would not survive to adulthood. As mortality falls, there is a time lag before families realize that more of their children are surviving; most of those eight kids, in other words, survived. Eventually parents adjust (through family planning and contraception) and have fewer children. See Dyson 2013.

17. Sharma 2016; Dyson 2013; Bloom and Williamson 1998; Bloom, Canning, and Sevilla 2003.

18. Beckley and Brands 2021; Fang 2016; Dyson 2013; Fong 2016.

19. On China's environmental damage, see Economy 2011; Jun Ma 2017.

20. Dasgupta et al. 2002.

21. Economy and Lieberthal 2007.

22. L. Wright 2022; Setser 2022.

23. Huang 2022; "Evergrande Is Not the Only Looming Danger in China's Financial System," *The Economist*, November 13, 2021.

24. James Kynge et al., "China Reckons with Its First Overseas Debt Crisis," *Financial Times*, July 21, 2022; Bennon and Fukuyama 2023.

25. Demographic challenges were exacerbated by the effects of access to sonogram technology and a cultural preference for sons, which produced a skewed birth ratio. See Liyan Qi, and Ming Li, "The One-Child Policy Supercharged China's Economic Miracle. Now It's Paying the Price," *Wall Street Journal*, July 11 2024; Hudson and den Boer 2004.

26. Sciubba 2022; Kroeber 2020; Lardy 2019.

27. On environmental devastation and Japan's economic rise, see Esarey et al. 2020; Broadbent 1999.

28. Lardy 2024.

29. On the US engagement policy toward China, see Friedberg 2022; Campbell and Ratner 2018. On modernization theory, see Lipset 1959; Moore 1967. On the "king's dilemma," see Huntington 1968.

30. Levitsky and Way 2010; Joachim and Brzezinski 1965; McMillan and Zoido 2004.

31. On media control and political influence, see Grossman, Margalit, and Mitts 2022; also see Copps 2023; Kaufmann 2004.

32. Michael Beckley, "No One Should Want to See a Dictator Get Old," *New York Times*, August 15, 2023; Shirk 2022; Alperovitch 2022; Evan Gershkovich, Thomas Grove, Drew Hinshaw, and Joe Parkinson, "Putin, Isolated and Distrustful, Leans on Handful of Hard-Line Advisers," *Wall Street Journal*, December 23, 2022.

33. Meng 2020; Boix and Svolik 2013; Gandhi 2008b

34. T. Lee 2002; Lorch and Bunk 2017; Lorentzen 2014; Lorentzen 2013. Martin Dimitrov (2023) argues that authoritarian leaders create institutions for accessing information, which prolongs the life of the regime under certain conditions.

35. Gilboy and Heginbotham 2001, 30.

36. Evan Osnos, "China's Age of Malaise," *The New Yorker*, October 23, 2023.

37. Carl Minzner, "The 20th Party Congress Is Another Step in Xi's Rise," *East Asia Forum*, October 16, 2022; Ian Bremmer, "'Maximum Xi Jinping' Will Be Costly for China and Its People," *Nikkei Asia*, January 18, 2023; Rudd 2022; Shirk 2022; Shambaugh 2021, 281–82;

Economy 2018; Minzner 2018. On CCP lessons learned from the Eastern European experience, see Shambaugh 2008, chap. 4.

38. Quoted in Evan Osnos, "Cycles of History," *New Yorker*, October 31, 2022.

39. Pearson Rithmire, and Tsai 2022, 136.

40. Quoted in Shambaugh 2021, 283.

41. Shirk 2022, 186–87.

42. Yan and Huang 2017.

43. Saich 2021.

44. Buckley, Chris, and Keith Bradsher, "China's Communists to Private Business: You Heed Us, We'll Help You," *New York Times*, September 17, 2022.

45. Bramble 2021;; Lingling Wei, "Xi Jinping Aims to Rein in Chinese Capitalism, Hew to Mao's Socialist Vision," *Wall Street Journal*, September 20, 2021. On the domestic politics of the affair, see Lingling Wei, "China Blocked Jack Ma's Ant IPO After Investigation Revealed Likely Beneficiaries," *Wall St. Journal*, February 16, 2021.

46. Li Yuan, "Why China Didn't Invent ChatGPT," *New York Times*, February 17, 2023, https://www.nytimes.com/2023/02/17/business/china-chatgpt-microsoft-openai.html.

47. Will Oremus, "China Is Doing What the U.S. Can't Seem To: Regulate Its Tech Companies," *Washington Post*, July 28, 2021; Nan Li and John Darwin Van Fleet, "Ant's Road to Redemption: How the Fintech Giant Can Save Itself," *The China Project*, May 18, 2021; Collier 2022;Huang and Lardy 2021.

48. "China's Tech Crackdown Starts to Ease"; Graham Webster, "China's Tech Turnaround," *The Wire China*, July 21, 2024; Daisuke Wakabayashi and Claire Fu, "From Disciplinarian to Cheerleader: Why China Is Changing Its Tone on Business," *New York Times*, January 12, 2023.

49. Zheping Huang, Jane Zhang, and Sara Zheng. 2023. "What Comes Next as China's Tech Crackdown Winds Down," *Washington Post*, July 24, 2023.

50. Zeyi Yang, "Why the Chinese Government Is Sparing AI from Harsh Regulations—For Now," *MIT Technology Review*, April 9, 2024, https://www.technologyreview.com/2024/04/09/1091004/china-tech-regulation-harsh-zhang/

51. Quoted in Zeyi Yang, "Why the Chinese Government Is Sparing AI from Harsh Regulations—For Now," *MIT Technology Review*, April 9, 2004; Zhang 2024.

52. Many leaders lament that the US tech economy has been driven not by strategic need but by venture capital, and they believe that Washington should play a stronger role in directing innovation. See Schmidt and Bajraktari 2022.

53. Weiss 2014.

54. Cheung 2022, 265.

55. Segal 2011.

56. Shambaugh 2021, 287 Also see Lardy 2019; Yu 2019.

57. Shirk 2022, 185.

58. Yu 2014, 2019.

59. S. Kennedy 2022.

60. Harrison et al. 2019. An important refinement is Allen et al. 2022, which explores the relationship between government involvement and firm productivity and profitability: expanding the concept of state-owned enterprises beyond the public/private binary.

61. Shirk 2017.

62. Shirk 2022, 183–206.

63. Rudd 2022.

64. Chris Buckley and Adam Wu, "Ending Term Limits for China's Xi Is a Big Deal. Here's Why," *New York Times*, March 10, 2018.

65. Xu and Guo 2023; Shambaugh 2016; Shirk 2022.

66. "Document 9: A ChinaFile Translation"; also see Deane 2021.

67. Deane 2021.

68. Howell 2019; Deane 2021.

69. S. Cook 2017, 7. On the Uighurs, see Ramzy and Buckley 2019; Kaltman 2014.

70. Bob Dietz, "In China, Mainstream Media as Well as Dissidents Under Increasing Pressure," Committee to Protect Journalists, December 17, 2014, https://cpj.org/2014/12/China-mainstream-media-as-well-as-dissidents-under-incre/#more.

71. David Schlesinger, Anne Henochowicz, and Yaqiu Wang, "Why Xi Jinping's Media Controls Are 'Absolutely Unyielding.'" *Foreign Policy*, March 17, 2016.

72. Schlesinger, Henochowicz, and Wang, "Why Xi Jinping's Media Controls Are 'Absolutely Unyielding.'"

73. Brenda Goh, "Three Hours a Week: Play Time's Over for China's Young Video Gamers," *Reuters*, August 31, 2021.

74. Beth Timmins, "China's Media Cracks Down on 'Effeminate' Styles," *BBC.com*, September 2, 2021.

75. James Palmer, "Why China Is Cracking Down on Private Tutoring," *Foreign Policy*, July 28, 2021.

76. Chris Buckley, "Xi Jinping Thought Explained: A New Ideology for a New Era," *New York Times*, February 26, 2018.

77. "China Schools: Xi Jinping Thought Introduced into Curriculum," *BBC News*, August 25, 2021.

78. Tom Philipps, "Xi Jinping Thought to Be Taught in Chinese Universities," *Guardian*, October 27, 2017.

79. On Xi's counterproductive policies, see Bremmer, "Maximum Xi Jinping"; Shirk 2022; Rudd 2022; Shambaugh 2021; Magnus 2018; Economy 2018; Beckley and Brands 2022.

80. Lind 2024; Lind and Press 2025.

81. Montgomery 2014; Anderson and Press 2023.

82. Edel and Shullman 2021; Polyakova and Meserole 2019; Kendall-Taylor and Shullman 2018; Cooley 2015; Mackinnon 2011.

83. Y. Zheng 1993; Petracca and Xiong 1990; L. Liu and J. Liu 1989; Sautman 1992. In response to Xi Jinping's "neo-authoritarian" turn, see Xiao 2019.

84. Xiao 2019, 3.

85. "In the Age of AI," interview with Orville Schell, *Frontline*, PBS. Transcript at https://www.pbs.org/wgbh/frontline/film/in-the-age-of-ai/transcript/.

86. Bianchini et al. 2022.

87. On types of innovation, see Pavitt 1984; Woertzel et al. 2015.

88. For an argument about an authoritarian advantage in AI, see Beraja et al. 2023.

89. Chin and Lin 2022; Deibert 2023; Xu Xu 2021; Kendall-Taylor, Frantz, and Wright 2020.

90. Lind 2024.

91. Lind 2024; Tunsjø 2018.

92. Doshi 2021; Mearsheimer 2014. On Chinese influence operations and domestic political intervention, see Ohlin and Hollis 2021; Lind 2018; Brady 2009. On Taiwan, see Hass, Glaser, and Bush 2023; Mastro 2021.

93. On Chinese efforts to shape international order, see Rühlig 2022; Oud 2020; Rolland 2020; Mazarr, Heath, and Cevallos 2018; Brunnermeier, Doshi, and James 2018.

94. On "freedom to roam" see Mearsheimer 2014. On US grand strategy in a more constrained era, see Lind and Press 2020; Allison 2020; Lind and Wohlforth 2019.

95. Yu 2024.

96. Quoted in Repucci and Slipowitz 2022.

97. Frantz 2018; Levitsky and Way 2010; Mounk 2018b; Kendall-Taylor and Frantz 2014b.

98. Repucci and Slipowitz 2022.

99. Way 2023; Oud 2020; Gokhale 2020; Diamond 2019.

100. Dukalskis 2021; Kendall-Taylor and Shullman 2018; Polyakova and Meserole 2019. On the struggle between democracy and authoritarianism, see Levitsky and Way 2010; Levitsky and Ziblatt 2018; Guriev and Treisman 2022.

101. A watershed contribution is Weeks 2014. Security studies scholars have begun to explore the heterogeneity of autocratic policies of coup-proofing; see Talmadge 2015; Reiter 2020.

102. Social scientists have long debated the relative performance of liberal versus authoritarian institutions (summarized in Cheibub, Gandhi, and Vreeland 2010).

103. On this distinction, see Friedberg 1988.

104. See, in particular, Schmidt and Bajraktari 2022.

105. Huntington 1988; Strange 1987. Also see Drezner 2010; Bell 2010.

106. Bhidé 2008.

107. Blanchette and Hass 2025; Vagle and Brooks 2025; Beckley 2018b; Brooks and Wohlforth 2016.

108. Peter Cowhey, Orville Schell, and Susan L. Shirk, "Meeting the China Challenge: A New American Strategy for Technology Competition." 21st Century China Center, https://asiasociety.org/sites/default/files/inline-files/report_meeting-the-china-challenge_2020.pdf; Peter Cowhey and Susan Shirk, "The Danger of Exaggerating China's Technological Prowess," *Wall Street Journal*, January 8, 2021.

109. On the shift toward industrial policy, see Rotman 2023. On microinstitutions, see Weiss 2014; Segal 2011.

110. On weaponized interdependence, see Farrell and Newman 2019. On the US-led export controls against China, see Lind and Mastanduno 2025.

111. Quote from Mearsheimer 2014. Heginbotham et. al 2018; Cordesman 2013; Shambaugh 2003.

112. On battlefield effectiveness and its drivers, see Lyall 2020; Talmadge 2015; Biddle 2010; R. Brooks and Stanley 2007; Biddle and Long 2004.

113. Ben Noon and Chris Bassler, "How Chinese Strategists Think AI Will Power a Military Leap Ahead," *Defense One*, September 21, 2021.

114. Talmadge 2015; Quinlivan 1999; Bickford 1994.

115. Heath 2025; Bitzinger and Char 2019; Fravel 2020; Wuthnow and Saunders 2024.

116. Freedman 2022; Sacks 2022.

117. For discussion, see Guriev and Treisman 2022.

118. Choung 1998; Kim 1997.

119. Manyika and Spence 2023.

120. Beraja et al. 2023.

121. Cowhey and Shirk, "The Danger of Exaggerating China's Technological Prowess."

References

Abrahamsen, Rita, and Gerald Bareebe. 2021. "Uganda's Fraudulent Election." *Journal of Democracy* 32, no. 2: 90–104. https://doi.org/10.1353/jod.2021.0021.

Abuza, Zachary. 1996. "The Politics of Educational Diplomacy in Vietnam: Educational Exchanges under Doi Moi." *Asian Survey* 36, no. 6: 618–31. https://doi.org/10.2307/2645795.

Acemoglu, Daron. 2008. "Oligarchic Versus Democratic Societies." *Journal of the European Economic Association* 6, no. 1: 1–44.

Acemoglu, Daron, and James A. Robinson. 2012. *Why Nations Fail: The Origins of Power, Prosperity, and Poverty*. New York: Crown Publishing.

Acemoglu, Daron, and James A. Robinson. 2019. *The Narrow Corridor: States, Societies, and the Fate of Liberty*. New York: Penguin.

Acemoglu, Daron, Georgy Egorov, and Konstantin Sonin. 2008. "Coalition Formation in Non-Democracies." *Review of Economic Studies* 75, no. 4: 987–1009.

Acemoglu, Daron, Simon Johnson, and James A. Robinson. 2005. "Institutions as a Fundamental Cause of Long-Run Growth." In *Handbook of Economic Growth*, Vol. 1A, edited by Philippe Aghion and Steven N. Durlauf. Amsterdam: Elsevier: 385–472

Acemoglu, Daron, Simon Johnson, James A. Robinson, and Pierre Yared. 2008. "Income and Democracy." *American Economic Review* 98, no. 3: 808–42.

Acemoglu, Daron, Suresh Naidu, Pascual Restrepo, and James A Robinson. 2019. "Democracy Does Cause Growth." *Journal of Political Economy* 127, no. 1: 47–100.

Agénor, Pierre-Richard, Otaviano Canuto, and Michael Jelenic. 2012. "Avoiding Middle-Income Growth Traps." *Economic Premise*, no. 98. https://openknowledge.worldbank.org/handle/10986/16954.

Ahn, Elise S., John Dixon, and Larissa Chekmareva. 2018. "Looking at Kazakhstan's Higher Education Landscape: From Transition to Transformation Between 1920 and 2015." In *25 Years of Transformations of Higher Education Systems in Post-Soviet*

Countries: Reform and Continuity, edited by Jeroen Huisman, Anna Smolentseva, and Isak Froumin. Cham, Switzerland: Palgrave Macmillan.

Aiyar, Shekhar, Romain A. Duval, Damien Puy, Yiqun Wu, and Longmei Zhang. 2013. "Growth Slowdowns and the Middle-Income Trap." International Monetary Fund.Al-Rasheed, Madawi. 2021. *The Son King: Reform and Repression in Saudi Arabia*. Oxford: Oxford University Press.

Al-Rawi, Ahmed. 2021. "Disinformation under a Networked Authoritarian State: Saudi Trolls' Credibility Attacks against Jamal Khashoggi." *Open Information Science* 5, no. 1: 140–62.

Alesina, Alberto, and Lawrence H. Summers. 1993. "Central Bank Independence and Macroeconomic Performance: Some Comparative Evidence." *Journal of Money, Credit and Banking* 25, no. 2: 151–62.

Alexander, Marlon. 2013. "Shaking Off Fears of Censorship in Singapore: Youth Hold Out Hope." *Southeast Asian Press Alliance*, September 25. https://ifex.org/shaking-off-the-fear-of-state-censorship-in-singapore-youth-hold-out-hope/.

Allen, Franklin, Junhui Cai, Xian Gu, Jun Qian, Linda Zhao, and Wu Zhu. 2022. "Centralization or Decentralization? The Evolution of State-Ownership in China." Social Science Research Network. http://dx.doi.org/10.2139/ssrn.4283197

Allen, Gregory C. 2022. "Choking Off China's Access to the Future of AI." Washington, DC: Center for Strategic and International Studies. https://www.csis.org/analysis/choking-chinas-access-future-ai.

Allen, Gregory C, Georgia Abramson, Lennart Heim, and Sam Winter-Levy. 2025. "The United Arab Emirates' AI Ambitions." Center for Strategic & International Studies, January 24. https://www.csis.org/analysis/united-arab-emirates-ai-ambitions.

Allen, Robert C. 2001. "The Rise and Decline of the Soviet Economy." *Canadian Journal of Economics* 34, no. 4: 859–81.

Allison, Graham. 2017. *Destined for War: Can America and China Escape Thucydides's Trap?* New York: Houghton Mifflin Harcourt.

Allison, Graham. 2020. "The New Spheres of Influence: Sharing the Globe with Other Great Powers." *Foreign Affairs*, April. Accessed September 20, 2021. https://www.foreignaffairs.com/articles/united-states/2020-02-10/new-spheres-influence.

Allison, Graham, and Robert Blackwill. 2013. "Interview: Lee Kuan Yew on the Future of US-China Relations." *Atlantic*, March 5.

Allison, Graham, Kevin Klyman, Karina Barbesino, and Hugo Yen. 2021. "The Great Tech Rivalry: China vs the U.S." Cambridge, MA: Harvard Kennedy School.

Alperovitch, Dmitri. 2022. "The Dangers of Putin's Paranoia." *Foreign Affairs*, March 18. Accessed September 30, 2023. https://www.foreignaffairs.com/articles/russia-fsu/2022-03-18/dangers-putins-paranoia.

Amsden, Alice H. 2001. *The Rise of "The Rest": Challenges to the West from Late-Industrializing Economies*. Oxford: Oxford University Press.

Anceschi, Luca. 2015. "The Persistence of Media Control under Consolidated Authoritarianism: Containing Kazakhstan's Digital Media," *Demokratizatsiya* 23, no. 3: 277–95.

Anderson, Nicholas, and Daryl G. Press. 2023. "Projecting Land-Based Air Power into East Asia: The Struggle to Defeat A2AD." Paper presented at the 2023 International Studies Association Annual Convention, Montréal, Canada, March 15–18.

Ang, Yuen Yuen. 2016. *How China Escaped the Poverty Trap*. Ithaca, NY: Cornell University Press.

Ang, Yuen Yuen. 2020. *China's Gilded Age: The Paradox of Economic Boom and Vast Corruption*. Cambridge: Cambridge University Press.

Ang, Yuen Yuen. 2022. "The Problem with Zero: How Xi's Pandemic Policy Created a Crisis for the Regime." *Foreign Affairs*, December 2. https://www.foreignaffairs.com/china/problem-zero-xi-pandemic-policy-crisis.

Ang, Yuen Yuen, Nan Jia, Bo Yang, and Kenneth G. Huang. 2023. "China's Low-Productivity Innovation Drive: Evidence from Patents." *Comparative Political Studies* 57, no. 12. https://doi.org/10.1177/00104140231209960.

Archer, Candace C., Glen Biglaiser, and Karl DeRouen Jr. 2007. "Sovereign Bonds and the 'Democratic Advantage': Does Regime Type Affect Credit Rating Agency Ratings in the Developing World?" *International Organization* 61, no 2: 341–65.

Arezki, Rabah. 2024. "Saudia Arabia's Economic Shifts Under MBS Raise Stability Concerns." Atlantic Council, February 26. https://www.atlanticcouncil.org/in-depth-research-reports/books/saudi-arabias-economic-shifts-under-mbs-raise-stability-concerns/.

Arias, Eric, James R. Hollyer, and B. Peter Rosendorff. 2018. "Cooperative Autocracies: Leader Survival, Creditworthiness, and Bilateral Investment Treaties." *American Journal of Political Science* 62, no. 4: 905–21. http://www.jstor.org/stable/26598791.

Aristotle. 2013. *Politics*. 2nd ed. Chicago: University of Chicago Press.

Armstrong, Scott. 1984. "Innovation Behind the Iron Curtain: The Problems and Promise of Soviet R&D." *Christian Science Monitor*, July 31.

Arrow, Kenneth J. 1962. "Economic Welfare and the Allocation of Resources for Invention." In *The Rate and Direction of Inventive Activity: Economic and Social Factors*, edited by R. R. Nelson. Princeton, NJ: Princeton University Press.

Asunka J., S. Brierley, M. Golden, E. Kramon, and G. Ofosu. 2019. Electoral Fraud or Violence: The Effect of Observers on Party Manipulation Strategies. *British Journal of Political Science* 49, no. 1: 129–51.

Atkinson, Robert D. 2024. "China is Rapidly Becoming a Leading Innovator in Advanced Industries." Washington DC: Information Technology & Innovation Foundation. https://itif.org/publications/2024/09/16/china-is-rapidly-becoming-a-leading-innovator-in-advanced-industries/.

Atkinson, Robert D., and Caleb Foote. 2019. "Is China Catching Up to the United States in Innovation?" Washington, DC: Information Technology and Innovation Foundation. Accessed May 7, 2020. https://itif.org/publications/2019/04/08/china-catching-united-states-innovation.

Aven, Petr, Vladimir Nazarov, and Samvel Lazaryan. 2016. "Twilight of the Petrostate." *National Interest*, May 17. Accessed November 30, 2022. https://nationalinterest.org/feature/twilight-the-petrostate-16235.

Awokuse, Titus O., and Hong Yin. 2010. "Intellectual Property Rights Protection and the Surge in FDI in China." *Journal of Comparative Economics* 38, no. 2: 217–24.

Azzem, Ibrahim. 2021. "Will It Ever Be Anwar Ibrahim's Turn?" *Foreign Policy*, March 28. https://foreignpolicy.com/2021/03/28/anwar-ibrahim-malaysia -opposition-pm/?gclid=Cj0KCQiA99ybBhD9ARIsALvZavUP3yahdnoc8jWyW BO6eqMqWU4zW_SDB94BGPl-d_uKlf0b3aKIJu8aAgbhEALw_wcB.

Babiarz, Kimberly Singer, Karen Eggleston, Grant Miller, and Qiong Zhang. 2015. "An Exploration of China's Mortality Decline under Mao: A Provincial Analysis, 1950–80." *Population Studies* 69, no. 1: 39–56.

Bader, Jeffrey. 2018. "U.S.-China Relations: Is It Time to End the Engagement?" Washington, DC: Brookings Institution. https://www.brookings.edu/articles /u-s-china-relations-is-it-time-to-end-the-engagement/.

Bai, Chong-En, Chang-Tai Hsieh, and Zheng Michael Song. 2020. "Special Deals with Chinese Characteristics." *NBER Macroeconomics Annual* 34, no. 1: 341–79.

Barr, Michael D. 2003. "J.B. Jeyaretnam: Three Decades as Lee Kuan Yew's Bête Noir." *Journal of Contemporary Asia* 33, no. 3: 299–317.

Barro, Robert J. 1996. *Getting It Right: Markets and Choices in a Free Society*. Cambridge, MA: MIT Press.

Barro, Robert J., and David B. Gordon. 1983. "Rules, Discretion and Reputation in a Model of Monetary Policy." *Journal of Monetary Economics* 12, no. 1: 101–21.

Bastiaens, Ida, and Nita Rudra. 2018. "Liberal Authoritarian Country Examples: Jordan and Tunisia." In *Democracies in Peril: Taxation and Redistribution in Globalizing Economies*. Cambridge: Cambridge University Press.

Baum, Matthew A., and David A. Lake. 2003. "The Political Economy of Growth: Democracy and Human Capital." *American Journal of Political Science* 47, no. 2: 333–47.

Beaulieu, Emily, Gary W. Cox, and Sebastian Saiegh. 2012. "Sovereign Debt and Regime Type: Reconsidering the Democratic Advantage." *International Organization* 66, no. 4: 709–38.

Beck, Martin, and Simone Hüser. 2015. "Jordan and the 'Arab Spring': No Challenge, No Change?" *Middle East Critique* 24, no. 1: 83–97. https://doi.org /10.1080/19436149.2014.996996.

Beckley, Michael. 2011. "China's Century? Why America's Edge Will Endure." *International Security* 36, no. 3: 41–78.

Beckley, Michael. 2018a. "The Power of Nations: Measuring What Matters." *International Security* 43, no. 2: 7–44.

Beckley, Michael. 2018b. *Unrivaled: Why America Will Remain the World's Sole Superpower*. Ithaca, NY: Cornell University Press.

Beckley, Michael. 2021. "Conditional Convergence and the Rise of China: A Political Economy Approach to Understanding Global Power Transitions." *Journal of Global Security Studies* 6, no. 1: ogaa010.

Beckley, Michael. 2022. *Danger Zone: The Coming Conflict with China*. New York: W. W. Norton.

Beckley, Michael, and Hal Brands. 2021. "The End of China's Rise: Beijing Is Running Out of Time to Remake the World." *Foreign Affairs*, October 1. Accessed November 23, 2021. https://www.foreignaffairs.com/articles /china/2021–10–01/end-chinas-rise.

Beckley, Michael, and Hal Brands. 2022. *Danger Zone: The Coming Conflict with China*. New York: W. W. Norton.

Beckley, Michael, Yusaku Horiuchi, and Jennifer M. Miller. 2018. "America's Role in the Making of the Japanese Miracle." *Journal of East Asian Studies* 18, no. 1: 1–21.

Bell, Daniel. 1976. *The Coming of Post-Industrial Society*. New York: Basic Books.

Bell, David A. 2010. "Political Columnists Think America Is in Decline. Big Surprise." *New Republic*, October 7.

Bellows, Heather E. 1993. "The Challenge of Informationalization in Post-Communist Societies." *Communist and Post-Communist Studies* 26, no. 2: 144–64.

Ben-Atar, Doron S. 2008. *Trade Secrets: Intellectual Piracy and the Origins of American Industrial Power*. New Haven, CT: Yale University Press.

Bennon, Michael, and Francis Fukuyama. 2023. "China's Road to Ruin." *Foreign Affairs*, August 22. Accessed November 5, 2023. https://www.foreignaffairs.com/china/belt-road-initiative-xi-imf.

Beraja, Martin, Andrew Kao, David Y. Yang, and Noam Yuchtman. 2023. "AI-Tocracy." *The Quarterly Journal of Economics* 138, no. 3: 1349–402.

Bergsten, C. Fred. 2022. *The United States vs. China: The Quest for Global Economic Leadership*. Cambridge: Polity.

Bernstein, T. P., and X. Lu. 2003. *Taxation without Representation in Rural China*. Cambridge: Cambridge University Press.

Bhagwati, Jagdish. 1966. *The Economics of Underdeveloped Countries*. New York: McGraw-Hill.

Bhagwati, Jagdish. 1978. *Anatomy and Consequences of Exchange Control Regimes*. Cambridge, MA: Ballinger.

Bhidé, Amar. 2008. *The Venturesome Economy: How Innovation Sustains Prosperity in a More Connected World*. Princeton: Princeton University Press.

Bian, Yanjie. 1994. "Guanxi and the Allocation of Urban Jobs in China." *China Quarterly*, no. 140: 971–99.

Bianchini, Stefano, Moritz Müller, and Pierre Pelletier. 2022. "Artificial Intelligence in Science: An Emerging General Method of Invention." *Research Policy* 51, no. 10: 104604.

Bickford, Thomas J. 1994. "The Chinese Military and Its Business Operations: The PLA as Entrepreneur." *Asian Survey* 34, no. 5: 460–74. https://doi.org/10.2307/2645058.

Biddle, Stephen. 2010. *Military Power: Explaining Victory and Defeat in Modern Battle. Military Power*. Princeton: Princeton University Press.

Biddle, Stephen, and Stephen Long. 2004. "Democracy and Military Effectiveness: A Deeper Look." *Journal of Conflict Resolution* 48, no. 4: 525–46.

Biddle, Stephen, and Ivan Oelrich. 2016. "Future Warfare in the Western Pacific: Chinese Antiaccess/Area Denial, U.S. AirSea Battle, and Command of the Commons in East Asia." *International Security* 41, no. 1: 7–48.

Bitzinger, Richard A., and James Char. 2019. *Reshapig the Chinese Military*. New York: Routledge.

Bjarnegård, Elin, and Pär Zetterberg. 2022. "How Autocrats Weaponize Women's Rights." *Journal of Democracy* 33, no. 2: 60–75.

Blainey, Geoffrey. 1988. *The Causes of War*. 3rd ed. Basingstoke: Macmillan.

Blanchette, Jude and Ryan Hass. 2025. "Know Your Rival, Know Yourself: Rightsizing the China Challenge." *Foreign Affairs* 104, no. 1: 88–101. https://www.foreignaffairs.com/united-states/know-your-rival-know-yourself-china.

Bloom, David E., David Canning, and Jaypee Sevilla. 2003. *The Demographic Dividend: A New Perspective on the Economic Consequences of Population Change.* Santa Monica, CA: RAND.

Bloom, David E., and Jeffrey G. Williamson. 1998. "Demographic Transitions and Economic Miracles in Emerging Asia." *The World Bank Economic Review* 12, no. 3: 419–55.

Bloomberg. 2023. "Huawei's Mate 60 Pro Phone Shows Large Step Toward Made-In-China Parts." Bloomberg.com. September 7. https://www.bloomberg.com/news/articles/2023-09-07/huawei-s-mate-60-shows-large-step-toward-made-in-china-parts?embedded-checkout=true.

Bogart, Dan, and Gary Richardson. 2011. "Property Rights and Parliament in Industrializing Britain." *Journal of Law & Economics* 54, no. 2: 241–74.

Boix, Carles, and Milan W. Svolik. 2013. "The Foundations of Limited Authoritarian Government: Institutions, Commitment, and Power-Sharing in Dictatorships." *Journal of Politics* 75, no. 2: 300–16.

Boldrin, Michele, and David K. Levine. 2013. "The Case Against Patents." *Journal of Economic Perspectives* 27, no. 1: 3–22. https//doi.org/10.1257/jep.27.1.3.

Boot, Max. 2006. *War Made New: Technology, Warfare, and the Course of History, 1500 to Today.* New York: Penguin.

Borthwick, Sally. 1978. "Schooling and Society in Late Qing China." Ph.D. diss., Australian National University. https://openresearch-repository.anu.edu.au/bitstream/1885/132453/4/b11747596_Borthwick_Sally.pdf.

Boyd, Carl and Akihiko Yoshida. 1995. *The Japanese Submarine Forces and World War II.* Annapolis, MD: Naval Institute Press.

Brady, Anne-Marie. 2009. *Marketing Dictatorship: Propaganda and Thought Work in Contemporary China.* Lanham, MD: Rowman & Littlefield.

Brainard, Jeffrey, and Dennis Normile. 2022. "China Rises to First Place in Most Cited Papers." *Science* 377, no. 6608: 799.

Bramble, Josh. 2021. "Beijing's Tech Sector Crackdown Sends a Clear Signal to Companies Going Global." Center for Strategic and International Studies, October 4. https://www.csis.org/blogs/new-perspectives-asia/beijings-tech-sector-crackdown-sends-clear-warning-companies-going.

Brands, Hal. 2018. "Democracy vs Authoritarianism: How Ideology Shapes Great-Power Conflict." *Survival* 60, no. 5: 61–114.

Brands, Hal, and Michael Beckley. 2021. "China Is a Declining Power—and That's the Problem." *Foreign Policy*, September 24. Accessed March 25, 2022. https://foreignpolicy.com/2021/09/24/china-great-power-united-states/.

Brands, Hal, and John Lewis Gaddis. 2021. "The New Cold War: America, China, and the Echoes of History." *Foreign Affairs*, October 19.

Brandt, Loren, and Thomas G. Rawski. 2008. "China's Great Economic Transformation." In *China's Great Transformation*, edited by Loren Brandt and Thomas G. Rawski. Cambridge: Cambridge University Press.

Branstetter, Lee, and Nicholas R. Lardy. 2008. "China's Embrace of Globalization." In *China's Great Economic Transformation*, edited by Loren Brandt and Thomas G. Rawski. Cambridge: Cambridge University Press.

Branstetter, Lee G., Guangwei Li, and Mengjia Ren. 2023. "Picking Winners? Government Subsidies and Firm Productivity in China." *Journal of Comparative Economics* 51, no. 4: 1186–99.

Bremmer, Ian. 2006. *The J Curve: A New Way to Understand Why Nations Rise and Fall*. New York: Simon & Schuster.

Breznitz, Dan. 2007. *Innovation and the State: Political Choice and Strategies for Growth in Israel, Taiwan, and Ireland*. New Haven, CT: Yale University Press.

Breznitz, Dan, and Michael Murphree. 2011. *Run of the Red Queen: Government, Innovation, Globalization, and Economic Growth in China*. New Haven, CT: Yale University Press.

Broadbent, Jeffrey. 1999. *Environmental Politics in Japan: Networks of Power and Protest*. Cambridge: Cambridge University Press.

Brodie, Bernard, and Fawn McKay Brodie. 1973. *From Crossbow to H-Bomb*. Bloomington, IN:Indiana University Press.

Brooks, Risa. 2008. *Shaping Strategy: The Civil-Military Politics of Strategic Assessment*. Princeton, NJ: Princeton University Press.

Brooks, Risa, and Elizabeth Stanley, eds. 2007. *Creating Military Power: The Sources of Military Effectiveness*. Palo Alto, CA: Stanford University Press.

Brooks, Stephen G. 2005. *Producing Security: Multinational Corporations, Globalization, and the Changing Calculus of Conflict*. Princeton, NJ: Princeton University Press.

Brooks, Stephen G., G John Ikenberry, and William C. Wohlforth. 2012. "Don't Come Home, America." *International Security* 37, no. 3: 7–51.

Brooks, Stephen G., and William C. Wohlforth. 2000. "Power, Globalization, and the End of the Cold War: Reevaluating a Landmark Case for Ideas." *International Security* 25, no. 3: 5–53.

Brooks, Stephen G., and William C. Wohlforth. 2015. "The Rise and Fall of the Great Powers in the Twenty-First Century: China's Rise and the Fate of America's Global Position." *International Security* 40, no. 3: 7–53.

Brooks, Stephen G., and William C. Wohlforth. 2016. *America Abroad: The United States' Global Role in the 21st Century*. Oxford: Oxford University Press.

Brooks, Stephen G., and William C. Wohlforth. 2023. "The Myth of Multipolarity." *Foreign Affairs*, April 18. Accessed April 18, 2023. https://www.foreignaffairs.com/united-states/china-multipolarity-myth.

Brown, Kerry. 2017. *China's World: The Foreign Policy of the World's Newest Superpower*. London: Bloomsbury.

Brummer, Matthew. 2020. "Innovation and Threats." *Defence and Peace Economics* 33, no. 5: 563–84. https://doi.org/1080/10242694.2020.1853984.

Brummer, Matthew and Jennifer Lind. 2022. "Artificial Intelligence and the Balance of Power: National Competitiveness in the Fourth Industrial Revolution." Paper Prepared for the Annual Meeting of the International Studies Association, Nashville, TN.

Brunnermeier, Markus, Rush Doshi, and Harold James. 2018. "Beijing's Bismarckian Ghosts: How Great Powers Compete Economically." *Washington Quarterly* 41, no. 3: 161–76.

Buchholz, Scott, Adam Routh, Joe Mariani, Akash Keyal, and Pankaj Kamleshkumar Kishnani. 2020. "The Realist's Guide to Quantum Technology and National Security." *Deloitte Review*, no. 27. Accessed July 1, 2024. https://

www2.deloitte.com/us/en/insights/industry/public-sector/the-impact-of -quantum-technology-on-national-security.html.

Bueno de Mesquita, Bruce, and George W. Downs. 2005. "Development and Democracy." *Foreign Affairs* 84, no. 5: 77–86. https://doi.org/10.2307/20031707.

Bueno de Mesquita, Bruce, J. D. Morrow, Randall M. Siverson, and Alastair Smith. 1999. "An Institutional Explanation of the Democratic Peace." *American Political Science Review* 93, no. 4: 791–807.

Bueno de Mesquita, Bruce, and Alastair Smith. 2011. *The Dictator's Handbook: Why Bad Behavior Is Almost Always Good Politics*. New York: PublicAffairs.

Bueno de Mesquita, Bruce, Alastair Smith, James D. Morrow, and Randolph M. Siverson. 2003. *The Logic of Political Survival*. Cambridge, MA: MIT Press.

Burns, John P. 1989. "Chinese Civil Service Reform: The 13th Party Congress Proposals." *China Quarterly*, no. 120: 739–70. http://www.jstor.org/stable/654556.

Burns, John P. 2003. "'Downsizing' the Chinese State: Government Retrenchment in the 1990s." *China Quarterly*, no. 175: 775–802.

Bush, Sarah Sunn. 2011. "International Politics and the Spread of Quotas for Women in Legislatures." *International Organization* 65, no. 1: 103–37.

Bush, Sarah Sunn, and Pär Zetterberg. 2021. "Gender Quotas and International Reputation." *American Journal of Political Science* 65: 326–41.

Bush, Vannevar. 1945. "Science: The Endless Frontier." Report to the President by Vannevar Bush, Director of the Office of Scientific Research and Development, July 1945. https://www.nsf.gov/od/lpa/nsf50/vbush1945.htm.

Byman, Daniel, and Jennifer Lind. 2010. "Pyongyang's Survival Strategy: Tools of Authoritarian Control in North Korea." *International Security* 35, no. 1: 44–74.

Caballero, Richard J., Takeo Hoshi, and Anil K. Kashyap. 2008. "Zombie Lending and Depressed Restructuring in Japan." *American Economic Review* 98, no. 5: 1943–77.

Cai, Jane. 2024. "Xi Jinping Calls for Scientists to Step Up Innovation in Hi-tech 'Battlefield'." *South China Morning Post*, June 25. https://www.scmp.com /news/china/politics/article/3267982/xi-jinping-calls-chinese-scientists-step -innovation-hi-tech-battlefield.

Campbell, Kurt M., and Ely Ratner. 2018. "The China Reckoning: How Beijing Defied American Expectations." *Foreign Affairs* 97, no. 2. Accessed January 29, 2020. https://www.foreignaffairs.com/articles/china/2018–02–13/china-reckoning.

Carson, Austin. 2018. *Secret Wars: Covert Conflict in International Politics*. Princeton University Press.

Castro, Daniel, Michael McLaughlin, and Eline Chivot. 2019. "Who Is Winning the AI Race? China, the EU or the United States?" Center for Data Innovation. https://www2.datainnovation.org/2019-china-eu-us-ai.pdf.

Chan, Joseph Man, and Jack Linchuan Qiu. 2003. "China." In *Media Reform: Democratizing the Media, Democratizing the State*, edited by Monroe E. Price, Beata Rozumilowicz, and Stefaan G. Verhulst. London: Routledge.

Chandler, Alfred D. 1977. *The Visible Hand: The Managerial Revolution in American Business*. Cambridge, MA: Harvard University Press.

Chandra, Siddharth, and Nita Rudra. 2015. "Reassessing the Links between Regime Type and Economic Performance: Why Some Authoritarian Regimes

Show Stable Growth and Others Do Not." *British Journal of Political Science* 45, no. 2: 253–85.

Chatterjee Miller, Manjari. 2021. *Why Nations Rise: Narrative and the Path to Great Power*. Oxford: Oxford University Press.

Che, Jiahua, and Yingyi Qian. 1998. "Insecure Property Rights and Government Ownership of Firms." *Quarterly Journal of Economics* 113, no. 2: 467–96.

Chee, Chan Heng. 1971. *Singapore—The Politics of Survival 1965–67*. Singapore: Oxford University Press.

Cheibub, José Antonio, Jennifer Gandhi, and James Raymond Vreeland. 2010. "Democracy and Dictatorship Revisited." *Public Choice* 143, no. 1/2: 67–101.

Chen, Hsi-en. 1980. *Chinese Education Since 1949: Academic and Revolutionary Models*. New York: Pergamon Press.

Chen, Jie, and Bruce J. Dickson. 2010. *Allies of the State: China's Private Entrepreneurs and Democratic Change*. Cambridge, MA: Harvard University Press.

Cheng, Siok-Hwa. 1977. "Singapore Women: Legal Status, Educational Attainment, and Employment Patterns." *Asian Survey* 17, no. 4: 358–74.

Cherry, Janet. 2010. "Civil Society and Development" In *The Oxford International Encyclopedia of Peace*, edited by Nigel J. Young. Oxford University Press.

Cheung, Tai Ming. 2022. *Innovate to Dominate: The Rise of the Chinese Techno-Security State*. Ithaca, NY: Cornell University Press.

Chin, Josh, and Liza Lin. 2022. *Surveillance State: Inside China's Quest to Launch a New Era of Social Control*. New York: St. Martin's Press.

"China's New Marriage Law." 1981. *Population and Development Review* 7, no. 2: 369–72. https://doi.org/10.2307/1972649.

Chorzempa, Martin. 2018. "How China Leapfrogged Ahead of the United States in the Fintech Race." Petersen Institute for International Economics. Accessed October 17, 2022. https://www.piie.com/blogs/china-economic-watch/how-china-leapfrogged-ahead-united-states-fintech-race.

Chorzempa, Martin. 2022. *The Cashless Revolution: China's Reinvention of Money and the End of America's Domination of Finance and Technology*. New York: PublicAffairs.

Choung, Jae-Yong. 1998. "Patterns of Innovation in Korea and Taiwan." *IEEE Transactions on Engineering Management* 45, no. 4: 357–65.

Christensen, Thomas J. 2015. *The China Challenge: Shaping the Choices of a Rising Power*. New York: W. W. Norton.

Clarke, Donald, Peter Murrell, and Susan Whiting. 2008. "The Role of Law in China's Economic Development." In *China's Great Economic Transformation*, edited by Loren Brandt and Thomas G. Rawski. Cambridge: Cambridge University Press.

Clay, Ian, and Robert D. Atkinson. 2022. "Wake Up, America: China Is Overtaking the United States in Innovation Capacity." Washington, DC: Information Technology & Innovation Foundation. Accessed January 23, 2023. https://itif.org/publications/2023/01/23/wake-up-america-china-is-overtaking-the-united-states-in-innovation-capacity/.

Cleverley, John. 1991. *The Schooling of China: Tradition and Modernity in Chinese Education*. 2nd ed. Sydney: Allen & Unwin.

Cohen, Eliot A. 1996. "A Revolution in Warfare." *Foreign Affairs* 75, no. 2: 37–54.

Colby, Elbridge A. 2021. *The Strategy of Denial: American Defense in an Age of Great Power Conflict*. New Haven, CT: Yale University Press.

Collier, Andrew. 2022. *China's Technology War: Why Beijing Took Down Its Tech Giants*. Singapore: Springer Nature.

Cook, Alexander C. 2016. *The Cultural Revolution on Trial: Justice in the Post-Mao Transition*. Cambridge: Cambridge University Press.

Cook, Sarah. 2017. *The Battle for China's Spirit: Religious Revival, Repression, and Resistance under Xi Jinping*. Lanham, MD: Rowman & Littlefield.

Cooley, Alexander. 2015. "Authoritarianism Goes Global: Countering Democratic Norms." *Journal of Democracy* 26, no. 3: 49–63.

Cooley, Alexander, and Jack Snyder. 2015. *Ranking the World*. Cambridge University Press.

Copeland, Dale C. 2000. *The Origins of Major War*. Ithaca, NY: Cornell University Press.

Copps, Michael. 2023. "Another Knife in the News." *Common Cause*, April 18. https://www.commoncause.org/articles/another-knife-in-the-news/.

Cornell, Svante E., S. Frederick Starr, and Albert Barro. 2021. "Political and Economic Reforms in Kazakhstan Under President Tokayev." Institute for Security and Development Policy. https://isdp.eu/publication/political-and-economic-reforms-in-kazakhstan-under-president-tokayev//.

Cortada, James W. 2012. *The Digital Flood: The Diffusion of Information Technology Across the U.S., Europe, and Asia*. Oxford University Press. Accessed January 19, 2023. https://doi.org/10.1093/acprof:oso/9780199921553.003.0006.

Costello, John, and Joe McReynolds. 2018. "China's Strategic Support Force: A Force for a New Era." *China Strategic Perspectives*, no. 13. Washington, DC: National Defense University.

Cowen, Tyler, and Alex Tabarrok. 2015. *Modern Principles: Macroeconomics*. 3rd ed. New York: Macmillan Learning.

Daemmrich, Arthur. 2017. "Invention, Innovation Systems, and the Fourth Industrial Revolution." *Technology & Innovation* 18, no. 4: 257–65.

Dai, Xinyuan. 2002. "Political Regimes and International Trade: The Democratic Difference Revisited." *American Political Science Review* 96, no. 1: 159–65.

Dang, Jianwei and Kazuyuki Motohashi. 2015. "Patent Statistics: A Good Indicator for Innovation in China? Patent Subsidy Program Impacts on Patent Quality." *China Economic Review* 35: 137–55.

Dasgupta, Susmita, Benoit Laplante, Hua Wang, and David Wheeler. 2002. "Confronting the Environmental Kuznets Curve." *Journal of Economic Perspectives* 16, no. 1: 147–68.

Davenport, Christian. 2007. "State Repression and Political Order." *Annual Review of Political Science* 10, no. 1: 1–22.

De Bary, William Theodore, and Richard John Lufrano. 1999. *Sources of Chinese Tradition*. New York: Columbia University Press.

De Smet, Aaron, Richard Steele, and Haimeng Zhang. 2021. "Shattering the Status Quo: A Conversation with Haier's Zhang Ruimin." *McKinsey Quarterly*, July 27. https://www.mckinsey.com/capabilities/people-and-organizational-performance/our-insights/shattering-the-status-quo-a-conversation-with-haiers-zhang-ruimin.

Deane, Lawrence. 2021. "Will There Be a Civil Society in the Xi Jinping Era? Advocacy and Non-Profit Organising in the New Regime." *Made in China*

Journal, July 15. Accessed April 28, 2022. https://madeinchinajournal .com/2021/07/15/will-there-be-a-civil-society-in-the-xi-jinping-era-advocacy -and-non-profit-organising-in-the-new-regime/.

Decker, Audrey. 2024. "Chinese Satellites Are Breaking the U.S. 'Monopoly' on Long-Range Targeting." *Defense One*, May 2. https://www.defenseone.com /threats/2024/05/new-chinese-satellites-ending-us-monopoly-ability-track -and-hit-long-distance-targets/396272/.

Deibert, Ronald J. 2023. "The Autocrat in Your iPhone." *Foreign Affairs*, December 12. Accessed December 20, 2022. https://www.foreignaffairs.com/world/autocrat -in-your-iphone-mercenary-spyware-ronald-deibert.

Deibert, R., Palfrey, J., Rohozinski, R., & Zittrain, J. 2011. *Access Contested: Security, Identity and Resistance in Asian Cyberspace*. Cambridge, MA: MIT Press.

Derby, Lauren H. 2009. *The Dictator's Seduction: Politics and the Popular Imagination in the Era of Trujillo*. Duke University Press.

Dernis, Hélène, and Mosahid Khan. 2004. "Triadic Patent Families Methodology." OECD Science, Technology and Industry Working Papers, No. 2004/02. OECD Publishing. Accessed November 22, 2022. https://doi.org/10.1787/443844125004.

Di Stefano, Giada, Alfonso Gambardella, and Gianmario Verona. 2012. "Technology Push and Demand Pull Perspectives in Innovation Studies: Current Findings and Future Research Directions." *Research Policy* 41, no. 8: 1283–95.

Diamond, Larry. 2002. "Elections without Democracy: Thinking About Hybrid Regimes." *Journal of Democracy* 13, no. 2: 21–35.

Diamond, Larry. 2019. *Ill Winds: Saving Democracy from Russian Rage, Chinese Ambition, and American Complacency*. New York: Random House.

Diamond, Larry, and Leonardo Morlino. 2004. "The Quality of Democracy: An Overview." *Journal of Democracy* 15, no. 4: 20–31.

Dickson, Bruce J. 2007. "Integrating Wealth and Power in China: The Communist Party's Embrace of the Private Sector." *China Quarterly*, no. 192: 827–54.

Dikötter, Frank. 2019. *How to Be a Dictator: The Cult of Personality in the Twentieth Century*. London: Bloomsbury.

Dimitrov, Martin K. 2017. "The Political Logic of Media Control in China." *Problems of Post-Communism* 64, no. 3–4: 121–27.

Dimitrov, Martin K. 2023. *Dictatorship and Information: Authoritarian Regime Resilience in Communist Europe and China*. Oxford: Oxford University Press.

Ding, Jeffrey. 2021. "China's Growing Influence over the Rules of the Digital Road." *Asia Policy* 28, no. 2: 33–42.

Ding, Jeffrey. 2023. "The Diffusion Deficit in Scientific and Technological Power: Re-Assessing China's Rise." *Review of International Political Economy* 31, no. 1: 173–98. https://doi.org/10.1080/09692290.2023.2173633.

Ding, Jeffrey. 2024. *Technology and the Rise of Great Powers: How Diffusion Shapes Economic Competition*. Princeton, NJ: Princeton University Press.

Dobson, William J. 2013. *The Dictator's Learning Curve: Inside the Global Battle for Democracy*. New York: Anchor.

"Document 9: A ChinaFile Translation—How Much Is a Hardline Party Directive Shaping China's Current Political Climate?" 2013. *ChinaFile*, November 8. www .chinafile.com/document-9-chinafile-translation.

Dollar, David, and Yiping Huang. 2022. *The Digital Financial Revolution in China*. Washington, DC: Brookings Institution Press.

Doner, Richard F. 2009. *The Politics of Uneven Development: Thailand's Economic Growth in Comparative Perspective*. Cambridge: Cambridge University Press.

Doner, Richard F., Allen Hicken, and Bryan K. Ritchie. 2009. "Political Challenges of Innovation in the Developing World 1." *Review of Policy Research* 26, no. 1-2: 151–71.

Doner, Richard F., Bryan K. Ritchie, and Dan Slater. 2005. "Systemic Vulnerability and the Origins of Developmental States: Northeast and Southeast Asia in Comparative Perspective." *International Organization* 59, no. 2: 327–61.

Doner, Richard F., and Ben Ross Schneider. 2016. "The Middle-Income Trap: More Politics than Economics." *World Politics* 68, no. 4: 608–44.

Donno, D., Fox, S., & Kaasik, J. 2022. International Incentives for Women's Rights in Dictatorships. *Comparative Political Studies* 55, no. 3: 451–92.

Dornbusch, Rudiger, and Sebastian Edwards, eds. 1991. *The Macroeconomics of Populism in Latin America*. Chicago: University of Chicago Press.

Doshi, Rush. 2020. "The United States, China, and the Contest for the Fourth Industrial Revolution." Prepared Statement before the US Senate Committee on Commerce, Science, and Transportation, Subcommittee on Security, July 30. https://www.commerce.senate.gov/services/files/6880BBA6-2AF0-4A43-8D32-6774E069B53E.

Doshi, Rush. 2021. *The Long Game: China's Grand Strategy and the Displacement of American Power*. Oxford: Oxford University Press.

Doyle, Michael W. 1986. "Liberalism and World Politics." *American Political Science Review* 80, no. 4: 1151–69.

Drezner, Daniel W. 2010. "The Declinism Industry in America." *Foreign Policy*, October 7.

Drezner, Daniel W. 2022. "The Death of the Democratic Advantage?" *International Studies Review* 24, no. 2: viac017.

Du, Julian, Yi Lu, and Zhigang Tao. 2008. "Economic Institutions and FDI Location Choice: Evidence from US Multinationals in China." *Journal of Comparative Economics* 36, no. 3: 412–29.

Duchâtel, Mathieu. 2021. "The Weak Links in China's Drive for Semiconductors." Paris: Institut Montaigne. Accessed July 11, 2023. https://www.institutmontaigne.org/en/publications/weak-links-chinas-drive-semiconductors.

Dukalskis, Alexander. 2021. *Making the World Safe for Dictatorship*. Oxford: Oxford University Press.

Dychtwald, Zak. 2021. "China's New Innovation Advantage." *Harvard Business Review* 99, no. 33: 55–60.

Dyson, Tim. 2013. *Population and Development: The Demographic Transition*. London: Zed Books.

Eaton, Jonathan, and Samuel Kortum. 1999. "International Technology Diffusion: Theory and Measurement." *International Economic Review* 40: 537 570. https://doi.org/10.1111/1468-2354.00028.

Eberstadt, Nicholas and Evan Abramsky. 2022. *The Changing Global Distribution of Highly Educated Manpower, 1950–2040: Findings and Implications*. Washington, DC: American Enterprise Institute.

Eckstein, Alexander. 1977. *China's Economic Revolution*. London: Cambridge University Press.

"Economic Diversification Efforts Paying Off in GCC Region but More Reforms Needed." 2023. World Bank. Press release, November 22. https://www .worldbank.org/en/news/press-release/2023/11/22/economic -diversification-efforts-paying-off-in-gcc-region-but-more-reforms-needed.

The Economist. 2008. "J.B. Jeyaretnam." October 9. https://www.economist.com /obituary/2008/10/09/jb-jeyaretnam.

Economy, Elizabeth. 2004. "Don't Break the Engagement." *Foreign Affairs*, May 1. Accessed January 25, 2023. https://www.foreignaffairs.com/articles/asia /2004-05-01/dont-break-engagement.

Economy, Elizabeth. 2011. *The River Runs Black: The Environmental Challenge to China's Future.* Ithaca, NY: Cornell University Press.

Economy, Elizabeth. 2018. *The Third Revolution: Xi Jinping and the New Chinese State.* Oxford: Oxford University Press.

Economy, Elizabeth, and Kenneth Lieberthal. 2007. "Scorched Earth: Will Environmental Risks in China Overwhelm its Opportunities?" *Harvard Business Review.*

Edel, Charles, and David O. Shullman. 2021. "How China Exports Authoritarianism." *Foreign Affairs*, September 16. Accessed March 22, 2023. https://www.foreignaffairs.com/articles/china/2021–09–16/how-china -exports-authoritarianism.

Egorov, Georgy, Sergei Guriev, and Konstantin Sonin. 2009. "Why Resource-Poor Dictators Allow Freer Media: A Theory and Evidence from Panel Data." *American Political Science Review* 103, no. 4: 645–68.

Eichengreen, Barry, Donghyun Park, and Kwanho Shin. 2013. "Growth Slowdowns Redux: New Evidence on the Middle-Income Trap." Working Paper No. 18673. National Bureau of Economic Research. Accessed January 22, 2020. http:// www.nber.org/papers/w18673.

Ermolaev, S. A. 2019. "Soviet Oil and Gas Dependence: Lessons for Contemporary Russia." *Problems of Economic Transition* 61, no. 10–12: 800–16.

Esarey, Ashley, Mary Alice Haddad, Joanna I. Lewis, and Stevan Harrell. 2020. *Greening East Asia: The Rise of the Eco-Developmental State.* Seattle: University of Washington Press.

Escribà-Folch, Abel, and Joseph Wright. 2015. *Foreign Pressure and the Politics of Autocratic Survival.* Oxford: Oxford University Press.

Evans, Peter. 1992. "The State as Problem and Solution: Predation, Embedded Autonomy, and Structural Change." In *The Politics of Economic Adjustment: International Constraints, Distributive Conflicts and the State*, edited by Stephan Haggard and Robert R. Kaufman. Princeton, NJ: Princeton University Press.

Evans, Peter, and James E. Rauch. 1999. "Bureaucracy and Growth: A Cross-National Analysis of the Effects of 'Weberian' State Structures on Economic Growth." *American Sociological Review* 64, no. 5: 748–65.

Fagerberg, Jan. 2006. "Innovation: A Guide to the Literature." In *The Oxford Handbook of Innovation*, edited by Jan Fagerberg, David Mowrey, and Richard R. Nelson. Oxford: Oxford University Press. https://smartech.gatech.edu /handle/1853/43180.

Fairbank, John King. 1971. *The United States and China.* 3rd ed. Harvard University Press.

Fakir, Intissar. 2021. "Consistency and Change: Morocco under King Mohammed VI." Middle East Institute. https://www.mei.edu/publications/consistency-and-change-morocco-under-king-mohammed-vi.

Fakir, Intissar, and Isabelle Werenfels. 2020. "In Morocco, Benevolent Authoritarianism Isn't Sustainable." Carnegie Endowment for International Peace. https://carnegieendowment.org/2020/07/29/in-morocco-benevolent-authoritarianism-isn-t-sustainable-pub-82395.

Fang, Cai. 2016. *China's Economic Growth Prospects: From Demographic Dividend to Reform Dividend*. Cheltenham, UK: Edward Elgar.

Fang, S., and E. Owen. 2011. "International Institutions and Credible Commitment of Non-Democracies." *Review of International Organizations* 6, 141–62.

Farrell, Henry, and Abraham L. Newman. 2019. "Weaponized Interdependence: How Global Economic Networks Shape State Coercion." *International Security* 44, no. 1: 42–79.

Fawcett, James T., and Siew-Ean Khoo. 1980. "Singapore: Rapid Fertility Transition in a Compact Society." *Population and Development Review* 6, no. 4: 549–79. https://doi.org/10.2307/1972926.

Fearon, James D. 1994. "Domestic Political Audiences and the Escalation of International Disputes." *American Political Science Review* 88, no. 3: 577–92.

Fedasiuk, Ryan, Alan Omar Loera Martinez, and Anna Puglisi. 2022. "A Competitive Era for China's Universities." Center for Security and Emerging Technology. Accessed September 29, 2022. https://cset.georgetown.edu/publication/a-competitive-era-for-chinas-universities/.

Fedderke, Johannes, and Dennis Jett. 2017. "What Price the Court of St. James? Political Influences on Ambassadorial Postings of the United States of America." *Governance* 30, no. 3: 483–515.

Fei, John C. H., and Gustav Ranis. 1964. *Development of the Labor Surplus Economy: Theory and Policy*. Homewood, IL: Irwin.

Feldstein, Steven. 2019. "The Road to Digital Unfreedom: How Artificial Intelligence Is Reshaping Repression." *Journal of Democracy* 30, no. 1, 40–52.

Feng, Wang. 2011. "The Future of a Demographic Overachiever: Long-Term Implications of the Demographic Transition in China." *Population and Development Review* 37: 173–90.

Fingleton, Eamon. 1995. *Blindside: Why Japan Is Still On Track to Overtake the United States by the Year 2000*. New York: Houghton Mifflin Harcourt.

Fong, Mei. 2016. *One Child: The Story of China's Most Radical Experiment*. New York: Oneworld.

Foote, Rosemary. 2000. *Rights Beyond Borders: The Global Community and the Struggle over Human Rights in China*. Oxford: Oxford University Press.

Foxley, Alejandro, and Sossdorf, Fernando. 2011. "Making the Transition: From Middle-Income to Advanced Economies." Carnegie Endowment for International Peace. https://carnegieendowment.org/research/2011/09/making-the-transition-from-middle-income-to-advanced-economies?lang=en.

Francis, Corinna-Barbara. 2001. "Quasi-Public, Quasi-Private Trends in Emerging Market Economies: The Case of China." *Comparative Politics* 33, no. 3: 275–94.

Frank, Andre Gunder. 1998. *ReORIENT: Global Economy in the Asian Age*. Berkeley: University of California Press.

Frantz, Erica. 2018. *Authoritarianism: What Everyone Needs to Know*. Oxford: Oxford University Press.

Frantz, Erica. 2019. "The Evolution of the Strongman." *Foreign Policy*, March 11.

Fravel, M. Taylor. 2020. *Active Defense: China's Military Strategy Since 1949*. Princeton, NJ: Princeton University Press.

Freedman, Lawrence. 2022. "Why War Fails." *Foreign Affairs* 101, no. 4. Accessed July 13, 2022. https://www.foreignaffairs.com/articles/russian-federation/2022–06–14/ukraine-war-russia-why-fails.

Friedberg, Aaron L. 1988. *The Weary Titan: Britain and the Experience of Relative Decline, 1895–1905*. Princeton, NJ: Princeton University Press.

Friedberg, Aaron L. 2015. *A Contest for Supremacy: China, America, and the Struggle for Mastery in Asia*. New York: W. W. Norton.

Friedberg, Aaron L. 2022. *Getting China Wrong*. New York: Wiley.

Frolic, Michael B. 1997. "State-Led Civil Society." In *Civil Society in China*, edited by Timothy Brook and Michael B. Frolic. Armonk, NY: East Gate Books.

Fuller, Douglas B. 2016. "China's Political Economy: Prospects for Technological Innovation-Based Growth." In *China's Innovation Challenge: Overcoming the Middle-Income Trap*, edited by Martin Kenney, Johann Peter Murmann, and Arie Y. Lewin. Cambridge: Cambridge University Press.

"Future of Productivity." 2015. Organization for Economic Cooperation and Development. https://doi.org/10.1787/9789264248533-en.

Galambos, Louis. 2012. *The Creative Society—and the Price Americans Paid for It*. New York: Cambridge University Press.

Gallagher, Mary E. 2004. "China: The Limits of Civil Society in a Late Leninist State." In *Civil Society and Political Change in Asia: Expanding and Contracting Democratic Space*, edited by Muthiah Alagappa. Palo Alto, CA: Stanford University Press.

Gallagher, Mary, and Jonathan Hanson. 2013. "Authoritarian Survival, Resilience, and the Selectorate Theory." In *Why Communism Did Not Collapse: Understanding Authoritarian Regime Resilience in Asia and Europe*, edited by Martin K. Dimitrov. Cambridge: Cambridge University Press.

Gallup, John Luke, Jeffrey D. Sachs, and Andrew D. Mellinger. 1999. "Geography and Economic Development." *International Regional Science Review* 22, no. 2: 179–232.

Galston, William. 2015. "Against Short-Termism: The Rise of Quarterly Capitalism Has Been Good for Wall Street—But Bad for Everyone Else." *Democracy*, no. 38.

Gandhi, Jennifer. 2008a. "Dictatorial Institutions and Their Impact on Economic Growth." *European Journal of Sociology* 49, no. 1: 3–30.

Gandhi, Jennifer. 2008b. *Political Institutions Under Dictatorship*. Cambridge: Cambridge University Press.

Gandhi, Jennifer, and Adam Przeworski. 2007. "Authoritarian Institutions and the Survival of Autocrats." *Comparative Political Studies* 40, no. 11: 1279–301.

Gardner, Tom. 2022. "I Was a War Reporter in Ethiopia. Then I Became the Enemy." *Economist*, June 24, 2022.

Gavin, Gabriel. 2024. "Georgia Defies EU and Backs Law Targeting Foreign Agents." *Politico*, May 14.

Geddes, Barbara. 1999. "What Do We Know About Democratization After Twenty Years?" *Annual Review of Political Science* 2, no. 1: 115–44.

Geddes, Barbara, Joseph Wright, and Erica Frantz. 2018. *How Dictatorships Work: Power, Personalization, and Collapse.* Cambridge: Cambridge University Press.

Gehlbach, Scott, and Philip Keefer. 2011. "Investment Without Democracy: Ruling-party Institutionalization and Credible Commitment in Autocracies." *Journal of Comparative Economics* 39, no. 2: 123–39.

Gehlbach, Scott, and Philip Keefer. 2012. "Private Investment and the Institutionalization of Collective Action in Autocracies: Ruling Parties and Legislatures." *The Journal of Politics* 74, no. 2: 621–35. https://doi.org/10.1017/s0022381611001952.

George, Cherian. 2007. "Consolidating Authoritarian Rule: Calibrated Coercion in Singapore." *Pacific Review* 20, no. 2: 124–45.

Gerring, John, Philip Bond, William T. Barndt, and Carola Moreno. 2005. "Democracy and Growth: A Historical Perspective." *World Politics* 57, no. 3: 323–64.

Gerschewski, Johannes. 2013. "The Three Pillars of Stability: Legitimation, Repression, and Co-Optation in Autocratic Regimes." *Democratization* 20, no. 1: 13–38.

Gewirtz, Julian. 2022. *Never Turn Back: China and the Forbidden History of the 1980s.* Cambridge, MA: Harvard University Press.

Gilboy, George, and Eric Heginbotham. 2001. "China's Coming Transformation." *Foreign Affairs* 80, no. 4: 26–39. https://doi.org/10.2307/20050224.

Gilboy, George J., and Benjamin L. Read. 2008. "Political and Social Reform in China: Alive and Walking." *Washington Quarterly* 31, no. 3: 143–61.

Giles, Martin. 2018. "The Man Turning China into A Quantum Superpower." MIT Technology Review, December 19. https://www.technologyreview.com/2018/12/19/1571/theman-turning-china-into-a-quantum-superpower/.

Gill, Indermit, and Homi Kharas. 2007. *An East Asian Renaissance : Ideas for Economic Growth.* Washington, DC: World Bank. Accessed September 11, 2022. https://openknowledge.worldbank.org/handle/10986/6798.

Gilli, Andrea, and Mauro Gilli. 2019. "Why China Has Not Caught Up Yet: Military-Technological Superiority and the Limits of Imitation, Reverse Engineering, and Cyber Espionage." *International Security* 43, no. 3: 141–89.

Gillis, Kay E. 2005. *Singapore Civil Society and British Power.* Singapore: Talisman.

Gilpin, Robert. 1981. *War and Change in World Politics.* Cambridge: Cambridge University Press.

Giuliano, Paola, Prachi Mishra, and Antonio Spilimbergo. 2013. "Democracy and Reforms: Evidence from a New Dataset." *American Economic Journal: Macroeconomics* 5, no. 4: 179–204.

Glaeser, Edward L., Rafael La Porta, Florencio Lopez-De-Silanes, and Andrei Shleifer. 2004. "Do Institutions Cause Growth?" *Journal of Economic Growth* 9: 271–303.

Glasius, Marlies. 2018a. "Extraterritorial Authoritarian Practices: A Framework." *Globalizations* 15, no. 2: 179–97.

Glasius, Marlies. 2018b. "What Authoritarianism Is . . . and Is Not: A Practice Perspective." *International Affairs* 94, no. 3: 515–33.

Goh, Chor Boon, and S. Gopinathan. 2008. "Education in Singapore: Developments Since 1965." In *An African Exploration of the East Asian Education Experience,* edited by Birger Fredriksen and Jee-Peng Tan. Washington, DC: World Bank.

Gokhale, Vijay. 2020. "China Is Gnawing at Democracy's Roots Worldwide," *Foreign Policy*, December 18. https://foreignpolicy.com/2020/12/18/china-democracy-ideology-communist-party.

Goldman, Merle, and Roderick MacFarquhar, eds. 1999. "Dynamic Economy, Declining Party-State." In *The Paradox of China's Post-Mao Reforms*. Cambridge, MA: Harvard University Press.

Goldstone, Jack A. 2012. "A Theory of Political Demography: Human and Institutional Reproduction." In *Political Demography: How Population Changes Are Reshaping International Security and National Politics*, edited by Jack A. Goldstone, Eric P. Kaufmann, and Monica Duffy Toft. Boulder, CO: Oxford University Press.

Goldstone, Jack A., Eric P. Kaufmann, and Monica Duffy Toft, eds. 2012. *Political Demography: How Population Changes Are Reshaping International Security and National Politics*. Oxford: Oxford University Press.

Gooroochurn, Nishaal, and Guntur Sugiyarto. 2005. "Competitiveness indicators in the travel and tourism industry." *Tourism Economics* 11, no. 1: 25–43.

Gourevitch, Peter Alexis. 1986. *Politics in Hard Times: Comparative Responses to International Economic Crises*. Ithaca, NY: Cornell University Press.

Graham, Loren R. 1998. *What Have We Learned about Science and Technology from the Russian Experience?* Palo Alto, CA: Stanford University Press.

Graham, Loren, ed. 1990. *Science and the Soviet Social Order:* Cambridge, MA: Harvard University Press.

Gray, Julia. 2009. "International Organization as a Seal of Approval: European Union Accession and Investor Risk." *American Journal of Political Science* 53, no. 4: 931–49.

Green, Elliott. 2011. "Patronage as Institutional Choice: Evidence from Rwanda and Uganda." *Comparative Politics* 43, no. 4: 421–38.

Greitens, Sheena Chestnut, Myunghee Lee, and Emir Yazici. 2019. "Counterterrorism and Preventive Repression: China's Changing Strategy in Xinjiang." *International Security* 44, no. 3: 9–47. https://doi.org/10.1162/isec_a_00368.

Grindle, Merilee S. 2004. *Despite the Odds: The Contentious Politics of Education Reform*. Princeton, NJ: Princeton University Press.

Groenewegen-Lau, Jeroen. 2024. "Whole-of-Nation Innovation: Does China's Socialist System Give It an Edge in Science and Technology?" Berlin: Mercator Institute for China Studies. Accessed August 14, 2024. https://merics.org/en/report/whole-nation-innovation-does-chinas-socialist-system-give-it-edge-science-and-technology.

Grossman, Guy, Yotam Margalit, and Tamar Mitts. 2022. "How the Ultrarich Use Media Ownership as a Political Investment." *Journal of Politics* 84, no. 4: 1913–31.

Guo, Baogang. 2003. "Political Legitimacy and China's Transition." *Journal of Chinese Political Science* 8, no. 1–2: 1–25. https://link.springer.com/article/10.1007/BF02876947.

Guo, Xuezhi. 2012. *China's Security State: Philosophy, Evolution, and Politics*. Cambridge: Cambridge University Press.

Guriev, Sergei, and Elias Papaioannou. 2022. "The Political Economy of Populism." *Journal of Economic Literature* 60, no. 3: 753–832. https://doi.org/10.1257/jel.20201595.

Guriev, Sergei, and Daniel Triesman. 2019. "Informational Autocrats." *Journal of Economic Perspectives* 33, no. 4: 100–27.

Guriev, Sergei, and Daniel Triesman. 2022. *Spin Dictators: The Changing Face of Tyranny in the 21st Century*. Princeton, NJ: Princeton University Press.

Gwartney, James D., Robert A Lawson, and Randall G Holcombe. 1999. "Economic Freedom and the Environment for Economic Growth." *Journal of Institutional and Theoretical Economics* 155, no. 4: 643–63.

Hadenius, Axel, and Jan Teorell. 2007. "Pathways from Authoritarianism." *Journal of Democracy* 18, no. 1: 143–56.

Haggard, Stephan. 1990. *Pathways from the Periphery: The Politics of Growth in the Newly Industrializing Countries*. Ithaca, NY: Cornell University Press.

Haggard, Stephan. 2000. "Interests, Institutions, and Policy Reform." In *Economic Policy Reform: The Second Stage*, edited by Anne O. Krueger. Chicago: University of Chicago Press.

Haggard, Stephan, and Robert R. Kaufman, eds. 1992. *The Politics of Economic Adjustment: International Constraints, Distributive Conflicts and the State*. Princeton, NJ: Princeton University Press.

Haggard, Stephan, and Robert Kaufman. 1995. *The Political Economy of Democratic Transitions*. Princeton, NJ: Princeton University Press.

Haggard, Stephan, Daniel Pinkston, and Jungkun Seo. 1999. "Reforming Korea Inc.: The Politics of Structural Adjustment Under Kim Dae Jung." *Asian Perspective* 23, no. 3: 201–35.

Hall, Stephen G. F., and Thomas Ambrosio. 2017. "Authoritarian Learning: A Conceptual Overview." *East European Politics* 33, no. 2: 143–61.

Hall, Bronwyn H., Adam Jaffe, and Manuel Trajtenberg. 2005. "Market Value and Patent Citations." *The RAND Journal of Economics* 36, no. 1 (2005): 16–38. http://www.jstor.org/stable/1593752.

Han, Rongbin, and Shao Li. 2022. "Scaling Authoritarian Information Control: How China Adjusts the Level of Online Censorship." *Political Research Quarterly* 75, no. 4: 1345–59.

Han, Shuangmiao, and Xin Xu. 2019. "How Far Has the State 'Stepped Back': An Exploratory Study of the Changing Governance of Higher Education in China (1978–2018). *Higher Education* 78: 931–46.

Hankla, Charles R., and Daniel Kuthy. 2013. "Economic Liberalism in Illiberal Regimes: Authoritarian Variation and the Political Economy of Trade." *International Studies Quarterly* 57, no. 3: 492–504.

Hannum, Emily, Jere Behrman, Meiyan Wang, and Jihong Liu. 2008. "Education in the Reform Era." In *China's Great Economic Transformation*, edited by Loren Brandt and Thomas G. Rawski. Cambridge University Press.

Hanson, Margaret, and Sarah Wilson Sokhey. 2021. "Higher Education as an Authoritarian Tool for Regime Survival: Evidence from Kazakhstan and around the World." *Problems of Post-Communism* 68, no. 3: 231–46.

Hanson, Philip. 2003. *The Rise and Fall of the Soviet Economy: An Economic History of the USSR 1945–1991*. Taylor & Francis.

Harhoff, Dietmar, Francis Narin, F. M. Scherer, and Katrin Vopel. 1999. "Citation Frequency and the Value of Patented Inventions." *Review of Economics and Statistics* 81, no. 3: 511–15.

Harrison, Ann, Marshall Meyer, Peichun Wang, Linda Zhao, and Minyuan Zhao. 2019. "Can a Tiger Change Its Stripes? Reform of Chinese State-Owned Enterprises in the Penumbra of the State." Working Paper No. 25475. National Bureau of Economic Research. Accessed December 20, 2022. https://www.nber.org/papers/w25475.

Hass, Ryan. 2021. "China Is Not Ten Feet Tall." *Foreign Affairs*, March 3. Accessed August 14, 2024. https://www.foreignaffairs.com/articles/china/2021-03-03/china-not-ten-feet-tall.

Hass, Ryan, Bonnie Glaser, and Richard Bush. 2023. *U.S.-Taiwan Relations: Will China's Challenge Lead to a Crisis?* Washington, DC: Brookings Institution Press.

Hassan, Tirana. 2023. "Repression Undermines Jordan's Reform Narrative." Human Rights Watch. https://www.hrw.org/news/2023/08/23/repression-undermines-jordans-reform-narrative.

Hayashi, Yuka. 2023. "US to Allow South Korean, Taiwan Chip Makers to Keep Operations in China." *Wall Street Journal*, June 12.

Hayek, Friedrich A. 1944. *The Road to Serfdom*. University of Chicago Press.

Heath, Timothy R. 2025. "The Chinese Military's Doubtful Combat Readiness." Santa Monica, CA: RAND Corporation. https://www.rand.org/pubs/perspectives/PEA830-1.html.

Heginbotham, Eric, Michael Nixon, Forrest E. Morgan, Jacob L. Heim, Jeff Hagen, Sheng Tao Li, Jeffrey Engstrom, Martin C. Libicki, et al. 2015. *The U.S.-China Military Scorecard: Forces, Geography, and the Evolving Balance of Power, 1996–2017.* Santa Monica, CA: RAND.

Heim, Irina, and Kairat Salimov. 2020. "The Effects of Oil Revenues on Kazakhstan's Economy." In *Kazakhstan's Diversification from the Natural Resources Sector*, edited by Irina Heim and Kairat Salimov. Cham, Switzerland: Springer Nature.

Henderson, J. Vernon, Zmarak Shalizi, and Anthony J. Venables. 2001. "Geography and Development." *Journal of Economic Geography* 1, no. 1: 81–105.

Herbst, Jeffrey. 2021. *The Politics of Reform in Ghana, 1982–1991*. Berkeley, CA: University of California Press.

Heston, Alan, and Terry Sicular. 2008. "China and Development Economics." In *China's Great Economic Transformation*, edited by Loren Brandt and Thomas G. Rawski. Cambridge: Cambridge University Press.

Heydemann, Steven, and Reinoud Leenders. 2011. "Authoritarian Learning and Authoritarian Resilience: Regime Responses to the 'Arab Awakening.'" *Globalizations* 8, no. 5: 647–53.

Hildebrandt, Timothy. 2013. "Self-Limiting Organizations and Codependent State–Society Relations: Environmental, HIV/AIDS, and Gay and Lesbian NGOs in China." In *Social Organizations and the Authoritarian State in China*, 1–22. Cambridge: Cambridge University Press.

Hirschman, Albert O. 1968. "The Political Economy of Import-Substituting Industrialization in Latin America." *Quarterly Journal of Economics* 82, no. 1: 1–32.

Hirschman, Albert O. 1991. *The Rhetoric of Reaction: Perversity, Futility, Jeopardy*. Cambridge, MA: Harvard University Press.

Ho, Elaine Wee. 2006. "Negotiating Belonging and Perceptions of Citizenship in a Transnational World: Singapore, a Cosmopolis?" *Social & Cultural Geography* 7, no. 3: 385–401.

Ho, Peter. 2001. "Who Owns China's Land? Policies, Property Rights and Deliberate Institutional Ambiguity." *China Quarterly*, no. 166: 394–421.

Hobson, John M. 2004. *The Eastern Origins of Western Civilisation*. Cambridge: Cambridge University Press.

Holm, David. 1991. *Art and Ideology in Revolutionary China*. Oxford: Clarendon University Press.

Horowitz, Michael C. 2010. *The Diffusion of Military Power*. Princeton, NJ: Princeton University Press.

Hou, Yue. 2019. *The Private Sector in Public Office*. Cambridge: Cambridge University Press.

"How Does Corruption Hinder China's Development?" 2018. ChinaPower Index. Center for Strategic and International Studies. https://chinapower.csis.org /china-corruption-development/.

"How Does Education in China Compare with Other Countries?" 2016. *ChinaPower* (blog), November 15. Updated August 26, 2020. https://chinapower.csis.org /education-in-china/.

Howell, Jude. 2019. "NGOs and Civil Society: The Politics of Crafting a Civic Welfare Infrastructure in the Hu–Wen Period." *China Quarterly*, no. 237: 58–81.

"How Saudi Arabia Is Reforming Education Through Technological Investment." 2022. Oxford Business Group. https://oxfordbusinessgroup.com/reports /saudi-arabia/2022-report/education-training/bright-future-a-raft-of -reforms-and-restructuring-of-the-education-system-boosts-technological -proficiency-and-equips-students-to-thrive/.

Hsiao, Mei-Chu W., and Frank S. T. Hsiao. 2004. "The Chaotic Attractor of Foreign Direct Investment—Why China? A Panel Data Analysis." *Journal of Asian Economics* 15: 641–70.

Hu, Albert G. Z., and Gary H. Jefferson. 2008. "Science and Technology in China." In *China's Great Economic Transformation*, edited by Loren Brandt and Thomas G. Rawski. Cambridge: Cambridge University Press.

Huang, Tianlei. 2022. "China's Looming Property Crisis Threatens Economic Stability." *Realtime Economics*, January 12. Petersen Institute for International Economics. Accessed February 17, 2022. https://www.piie.com/blogs /realtime-economic-issues-watch/chinas-looming-property-crisis-threatens -economic-stability.

Huang, Tianlei, and Nicholas R. Lardy. 2021. "Is the Sky Really Falling for Private Firms in China?" *China Economic Watch*, October 14. Petersen Institute for International Economics. https://www.piie.com/blogs/china-economic-watch /sky-really-falling-private-firms-china/.

Hudson, Valerie M., and Andrea den Boer. 2004. *Bare Branches: The Security Implications of Asia's Surplus Male Population*. Cambridge: MIT Press.

Huff, W. G. 1995. "The Developmental State, Government, and Singapore's Economic Development Since 1960." *World Development* 23, no. 8: 1421–38.

Huntington, Samuel P. 1968. *Political Order in Changing Societies*. New Haven, CT: Yale University Press.

Huntington, Samuel P. 1988. "The U.S.—Decline or Renewal?" *Foreign Affairs* 67, no. 2: 76–96.

Hyde, Susan D., and Elizabeth N. Saunders. 2020. "Recapturing Regime Type in International Relations: Leaders, Institutions, and Agency Space." *International Organization* 74, no. 2: 363–95.

Ichikawa, Hiroshi. 2020. *Soviet Science and Engineering in the Shadow of the Cold War*. New York: Routledge.

Ikenberry, G. John. 2000. *After Victory: Institutions, Strategic Restraint, and the Rebuilding of Order after Major Wars*. Princeton, NJ: Princeton University Press.

Ikenberry, G. John. 2011. "The Future of the Liberal World Order." *Foreign Affairs* 90, no. 3: 56–68.

Ikenberry, G. John. 2012. *Liberal Leviathan: The Origins, Crisis, and Transformation of the American World Order*. Princeton, NJ: Princeton University Press.

Imai, Kosuke, In Song Kim, and Erik H. Wang. 2022. "Matching Methods for Causal Inference with Time-Series Cross-Sectional Data." *American Journal of Political Science* 67, no. 3: 1–19.

Inglehart, Ronald, and Christian Welzel. 2009. "How Development Leads to Democracy: What We Know About Modernization." *Foreign Affairs* 88, no. 2: 33–48.

Ingram, George. 2020. "Civil Society: An Essential Ingredient of Development." Brookings Institution, April 6. https://www.brookings.edu/blog/up-front/2020/04/06/civil-society-an-essential-ingredient-of-development/.

International Monetary Fund. 2025. "World Economic Outlook Database: GDP per Capita, Current Prices: US dollars per Capita." https://www.imf.org/.

Jacques, Martin. 2009. *When China Rules the World: The End of the Western World and the Birth of a New Global Order*. New York: Penguin.

Jaffe, Adam B., Manuel Trajtenberg, and Michael S. Fogarty. 2000. "Knowledge Spillovers and Patent Citations: Evidence from a Survey of Inventors." *American Economic Review* 90, no. 2: 215–18. https://doi.org/10.1257/aer.90.2.215.

Janos, Andrew. 2000. *East Central Europe in the Modern World: The Politics of Borderlands from Pre- to Postcommunism*. Palo Alto, CA: Stanford University Press.

Jerit, Jennifer, and Yangzi Zhao. 2020. "Political Misinformation." *Annual Review of Political Science* 23, no. 1: 77–94.

Ji, Elliot. 2023. "Great Leap Nowhere: The Challenges of China's Semiconductor Industry." War on the Rocks, February 24. Accessed August 7, 2023. https://warontherocks.com/2023/02/great-leap-nowhere-the-challenges-of-chinas-semiconductor-industry/.

Joachim, Friedrich Carl, and Zbignew K. Brzezinski. 1965. *Totalitarian Dictatorship and Autocracy*. Cambridge, MA: Harvard University Press.

Johnson, Chalmers. 1982. *MITI and the Japanese Miracle: The Growth of Industrial Policy, 1925–1975*. Palo Alto, CA: Stanford University Press. http://www.sup.org/books/title/?id=2791.

Johnson, Kay Ann. 1985. *Women, the Family, and Peasant Revolution in China*. Chicago: University of Chicago Press.

Jones, Benjamin F., and Benjamin A. Olken. 2008. "The Anatomy of Start-Stop Growth." *Review of Economics and Statistics* 90, no. 3: 582–87.

Jones, Calvert W. 2017. *Bedouins Into Bourgeois: Remaking Citizens for Globalization*. Cambridge: Cambridge University Press.

Juhász, Réka, Nathan Lane, and Dani Rodrik. 2024. "The New Economics of Industrial Policy." *Annual Review of Economics* 16 (August 2024): 213–42.

Kadir, Netina Tan. 2020. "Digital Learning and Extending Electoral Authoritarianism in Singapore." *Democratization* 27, no. 6: 1073–91.

Kadir, Suzaina. 2004. "Singapore: Engagement and Autonomy Within the Political Status Quo." In *Civil Society and Political Change in Asia: Expanding and Contracting Democratic Space*, edited by Muthiah Alagappa. Palo Alto: Stanford University Press.

Kahn, Herman. 1970. *Challenge and Response*. Englewood Cliffs, NJ: Prentice-Hall.

Kakachia, Kornely, and Bidzina Lebanidze. 2023. "Georgia's Slide to Authoritarianism." *Carnegie Europe*. https://carnegieeurope.eu/strategiceurope/89260.

Kaltman, Blaine. 2014. *Under the Heel of the Dragon: Islam, Racism, Crime, and the Uighur in China*. Athens: Ohio University Press.

Kang, David C. 2002. *Crony Capitalism: Corruption and Development in South Korea and the Philippines*. Cambridge: Cambridge University Press.

Kania, Elsa. 2020. "'AI Weapons' in China's Military Innovation." In *Global China: Assessing China's Growing Role in the World*, edited by Tarun Chhabra, Rush Doshi, Ryan Hass, and Emilie Kimball. Washington, DC: Brookings Institution Press.

Kania, Elsa. 2021a. "Artificial Intelligence in China's Revolution in Military Affairs." *Journal of Strategic Studies* 44, no. 4: 515–42.

Kania, Elsa. 2021b. "China's Drive for Innovation within a World of Profound Changes." *Asia Policy* 16, no. 2: 17–31.

Kania, Elsa B., and John Costello. 2018. "Quantum Hegemony? China's Ambitions and the Challenge to U.S. Innovation Leadership." Washington, DC: Center for a New American Security. Accessed July 1, 2024. https://www.cnas.org/publications/reports/quantum-hegemony.

Kania, Elsa, and Adam Segal. 2021. "Globalized Innovation and Great Power Competition: The US-China Tech Clash." In *After Engagement: Dilemmas in US-China Security Relations*, edited by Jacques deLisle and Avery Goldstein. Washington, DC: Brookings Institution Press.

Kaplan, Robert D. 2023. *The Loom of Time: Between Empire and Anarchy from the Mediterranean to China*. New York: Random House.

Kastner, Jill, and William C. Wohlforth. 2021. "A Measure Short of War: The Return of Great-Power Subversion." *Foreign Affairs* 100, no. 4: 118–31.

Katz, Richard. 1998. *Japan: The System That Soured*. New York: Routledge.

Kaufman, Victor S. 1998. "A Response to Chaos: The United States, the Great Leap Forward, and the Cultural Revolution, 1961–1968." *Journal of American-East Asian Relations* 7, no. 1–2: 73–92.

Kaufmann, Chaim. 2004. "Threat Inflation and the Failure of the Marketplace of Ideas: The Selling of the Iraq War." *International Security* 29, no. 1: 5–48. https://doi.org/10.1162/0162288041762940.

Keefer, Philip, and Stephen Knack. 1997. "Why Don't Poor Countries Catch Up? A Cross-National Test of an Institutional Explanation." *Economic Inquiry* 35, no. 3: 590–602.

Keeler, John. 1993. "Opening the Window for Reform: Mandates, Crises, and Extraordinary Policy Making." *Comparative Political Studies* 25, no. 4: 433–86.

Kendall-Taylor, Andrea, and Erica Frantz. 2014a. "How Autocracies Fall. *Washington Quarterly* 37, no. 1: 35–47.

Kendall-Taylor, Andrea, and Erica Frantz. 2014b. "Mimicking Democracy to Prolong Autocracies." *Washington Quarterly* 37, no. 4: 71–84.

Kendall-Taylor, Andrea, Erica Frantz, and Joseph Wright. 2020. "The Digital Dictators: How Technology Strengthens Autocracy." *Foreign Affairs* 99, no. 2: 103–15.

Kendall-Taylor, Andrea, and David O. Shullman. 2018. "How Russia and China Undermine Democracy." *Foreign Affairs*, October 2. https://www.foreignaffairs.com/articles/china/2018-10-02/how-russia-and-china-undermine-democracy.

Kennedy, Andrew B. 2018. *The Conflicted Superpower: America's Collaboration with China and India in Global Innovation.* New York: Columbia University Press.

Kennedy, Andrew B. 2019. "China's Rise as a Science Power: Rapid Progress, Emerging Reforms, and the Challenge of Illiberal Innovation." *Asian Survey* 59, no. 6: 1022–43.

Kennedy, Paul. 1987. *The Rise and Fall of the Great Powers: Economic Change and Military Conflict from 1500 to 2000.* New York: Knopf.

Kennedy, Scott. 2022. "Data Dive: The Private Sector Drives Growth in China's High-Tech Exports." Center for Strategic and International Studies, April 28. https://www.csis.org/blogs/trustee-china-hand/data-dive-private-sector-drives-growth-chinas-high-tech-exports.

Kerkvliet, Benedict J. Tria. 2014. "Government Repression and Toleration of Dissidents in Contemporary Vietnam." In *Politics in Contemporary Vietnam: Party, State, and Authority Relations,* edited by Jonathan London. London: Palgrave Macmillan. Accessed April 8, 2022. https://www.degruyter.com/document/doi/10.7591/9781501736391/html.

Kerkvliet, Benedict J. Tria. 2019. *Speaking Out in Vietnam: Public Political Criticism in a Communist Party–Ruled Nation.* Ithaca, NY: Cornell University Press.

Kerr, William R., and Frederic Robert-Nicoud. 2020. "Tech Clusters." *Journal of Economic Perspectives* 34, no. 3: 50–76. DOI: 10.1257/jep.34.3.50.

Khan, Mohsin S., and Zuliu Hu. 1997. "Why Is China Growing So Fast?" *Economic Issues,* no. 8. https://www.imf.org/EXTERNAL/PUBS/FT/ISSUES8/INDEX.HTM.

Khanin, G. I. 2003. "The 1950s: The Triumph of the Soviet Economy." *Europe-Asia Studies* 55, no. 8: 1187–1211. http://www.jstor.org/stable/3594504.

Kharas, Homi, and Harinder Kohli. 2011. "What Is the Middle Income Trap, Why Do Countries Fall into It, and How Can It Be Avoided?" *Global Journal of Emerging Market Economies* 3, no. 3: 281–89.

Khrushchev, Nikita. 1974. *Khrushchev Remembers: The Last Testament.* Translated by Strobe Talbott. Boston: Little, Brown.

"'Kill the Chicken to Scare the Monkeys': Suppression of Free Expression and Assembly in Singapore." 2017. Human Rights Watch. https://www.hrw.org/report/2017/12/13/kill-chicken-scare-monkeys/suppression-free-expression-and-assembly-singapore#:~:text=They%20slaughter%20the%20chicken%20to,you%20will%20pay%20a%20price.

Kim, Linsu. 1997. *Imitation to Innovation: The Dynamics of Korea's Technological Learning.* Boston: Harvard Business School Press. http://hdl.handle.net/2027/heb00987.0001.001.

Kim, Wonik, and Jennifer Gandhi. 2010. "Coopting Workers under Dictatorship." *Journal of Politics* 72, no. 3: 646–58.

King, Gary, Jennifer Pan, and Margaret E. Roberts. 2013. "How Censorship in China Allows Government Criticism but Silences Collective Expression." *American Political Science Review* 107, no. 2: 326–43.

Kingstone, Peter R. 2010. *Crafting Coalitions for Reform: Business Preferences, Political Institutions, and Neoliberal Reform in Brazil.* State College, PA: Penn State University Press.

Kirby, William C. 2022. *Empires of Ideas: Creating the Modern University from Germany to America to China.* Cambridge, MA: Harvard University Press.

Kliman, Daniel M. 2015. *Fateful Transitions: How Democracies Manage Rising Powers, from the Eve of World War I to China's Ascendance.* Philadelphia, PA: University of Pennsylvania Press.

Kline, S. J., and Rosenberg, N. 1986. "An Overview of Innovation." In *The Positive Sum Strategy: Harnessing Technology for Economic Growth,* 275–307, edited by R. Landau and N. Rosenberg. Washington DC: National Academies Press.

Klotz, Audie. 2018. *Norms in International Relations: The Struggle Against Apartheid.* Ithaca, NY: Cornell University Press.

Koh, Gillian, and Debbie Soon. 2012. "The Future of Singapore's Civil Society." *Social Space,* 92–98. Lien Center for Social Innovation, Lee Kuan Yew School of Public Policy, https://lkyspp.nus.edu.sg/docs/default-source/ips/socialspace2012 -gilliankoh-debbiesoon.pdf.

Kohli, Atul. 2004. *State-Directed Development: Political Power and Industrialization in the Global Periphery.* New York: Cambridge University Press.

Kono, Daniel Yuichi. 2015. "Authoritarian Regimes." In *The Oxford Handbook of the Political Economy of International Trade,* edited by Lisa L. Martin. Oxford: Oxford University Press: 298–315.

Kortum, S., and J. Lerner. 2000. "Assessing the Impact of Venture Capital on Innovation." *RAND Journal of Economics* 31: 674–92.

Krastev, Ivan. 2011. "Paradoxes of the New Authoritarianism." *Journal of Democracy* 22, no. 2: 5–16.

Krekó, Péter. 2017. "Hungary: Crackdown on Civil Society á la Russe Continues." Center for Strategic and International Studies, May 18. https://www.csis.org /blogs/international-consortium-closing-civic-space/hungary-crackdown -civil-society-la-russe.

Kroeber, Arthur R. 2020. *China's Economy: What Everyone Needs to Know.* 2nd ed. Oxford: Oxford University Press.

Kroenig, Matthew. 2020. *The Return of Great Power Rivalry: Democracy versus Autocracy from the Ancient World to the U.S. and China.* Oxford: Oxford University Press.

Kroenig, Matthew, and Dan Negrea. 2024. *We Win, They Lose: Republican Foreign Policy and the New Cold War.* New York: Republic Book.

Krueger, Anne O. 1990. "Government Failures in Development." *Journal of Economic Perspectives* 4, no. 3: 9–23.

Krueger, Anne O. 1997. "Trade Policy and Economic Development: How We Learn." Working Paper No. 5896. National Bureau of Economic Research. https://www.nber.org/papers/w5896.

Krugman, Paul. 1994. "The Myth of Asia's Miracle." *Foreign Affairs* 73, no. 6: 62–78.

Krugman, Paul. 1995. *Development, Geography, and Economic Theory.* Cambridge: MIT Press.

Kuhn, Thomas S. 1962. *The Structure of Scientific Revolutions*. Chicago: University of Chicago Press.

Kum, Wai. 2008. "Fifty Years and More of the 'Women's Charter' of Singapore." *Singapore Journal of Legal Studies*, 1–24. http://www.jstor.org/stable/24869349.

Kurlantzick, Joshua. 2003. "The Dragon Still Has Teeth: How the West Winks at Chinese Repression." *World Policy Journal* 20, no. 1: 49–58.

"Kyrgyzstan: Unprecedented Crackdown on Civil Society Threatens Human Rights and the Country's International Standing." 2024. Amnesty International, February 8. https://www.amnesty.org/en/latest/news/2024/02/kyrgyzstan-unprecedented-crackdown-on-civil-society-threatens-human-rights-and-countrys-international-standing/.

Lamb, Gregory M. 1982. "High-Tech Copycats: They Take It Apart or Steal It. *Christian Science Monitor*, July 21.

Landes, David S. 1969. *The Unbound Prometheus: Technological Change and Industrial Development in Western Europe from 1750 to the Present*. New York: Cambridge University Press.

Landry, Pierre François. 2008. *Decentralized Authoritarianism in China: The Communist Party's Control of Local Elites in the Post-Mao Era*. Vol. 1. New York: Cambridge University Press.

Laothamatas, Anek. 1994. "From Clientelism to Partnership: Business-Government Relations in Thailand." In *Business and Government in Industrialising Asia*, edited by Andrew J MacIntyre. Ithaca, NY: Cornell University Press.

Lardy, Nicholas. 2019. *The State Strikes Back: The End of Economic Reform in China?* Washington, DC: Petersen Institute for International Economics.

Lardy, Nicholas. 2024. "China Is Still Rising." *Foreign Affairs*. Accessed April 14, 2024. https://www.foreignaffairs.com/united-states/china-still-rising.

Lascurettes, Kyle M. 2020. *Orders of Exclusion: Great Powers and the Strategic Sources of Foundational Rules in International Relations*. Oxford: Oxford University Press.

Lau, Lawrence J., Yingyi Qian, and Gerard Roland. 2008. "Reform Without Losers: An Interpretation of China's Dual-Track Approach to Transition." *Journal of Political Economy* 108, no. 1: 120–43.

Lauwerys, Joseph Albert, and Muhammed Shamsul Huq. 2024. "Education Under Communism." *Encyclopedia Brittanica*. Accessed June 22, 2024. https://www.britannica.com/topic/education/Education-under-communism.

Lavely, William, Xiao Zhenyu, Li Bohua, and Ronald Freedman. 1990. "The Rise in Female Education in China: National and Regional Patterns." *China Quarterly*, no. 121: 61–93.

Law, Wing-Wah. 2002. "Legislation, Education Reform and Social Transformation: The People's Republic of China's Experience." *International Journal of Educational Development* 22, no. 6: 579–602.

"Laws of Attrition: Crackdown on Russia's Civil Society After Putin's Return to the Presidency." 2013. Human Rights Watch. https://www.hrw.org/report/2013/04/24/laws-attrition/crackdown-russias-civil-society-after-putins-return-presidency.

Layne, Christopher. 2018. "The US–Chinese Power Shift and the End of the Pax Americana." *International Affairs* 94, no. 1: 89–111.

Lebovic, James H., and Erik Voeten. 2009. "The Cost of Shame: International Organizations and Foreign Aid in the Punishing of Human Rights Violators." *Journal of Peace Research* 46, no. 1: 79–97.

Lee, Jong-wha. 1997. "Economic Growth and Human Development in the Republic of Korea, 1945–1992." Occasional Paper No. 24. New York: UN Development Programme.

Lee, Kai-Fu. 2018. *AI Superpowers: China, Silicon Valley, and the New World Order*. Boston: Houghton Mifflin Harcourt.

Lee, Kuan Yew. 2000. *From Third World to First: The Singapore Story: 1965–2000*. New York: HarperCollins.

Lee, Terence. 2002. "The Politics of Civil Society in Singapore." *Asian Studies Review* 26, no. 1: 97–117.

Leicester, John. 2002. "Last Gasp for China's Steam Engines." *Washington Post*, July 7.

Lendon, Brad. 2018. "China's New Destroyers: 'Power, Prestige and Majesty.'" CNN.com, July 13. https://www.cnn.com/2018/07/13/asia/china-new-destroyers-intl/index.html.

Lendon, Brad. 2022. "Never Mind China's New Aircraft Carrier, These Are the Ships the US Should Worry About." CNN.com, June 26.

Levinger, Matthew Bernard. 2000. *Enlightened Nationalism: The Transformation of Prussian Political Culture, 1806–1848*. Oxford: Oxford University Press.

Levitsky, Steven, and Lucan A. Way. 2002. "Elections Without Democracy: The Rise of Competitive Authoritarianism." *Journal of Democracy* 13, no. 2: 51–65.

Levitsky, Steven, and Lucan A. Way. 2010. *Competitive Authoritarianism: Hybrid Regimes after the Cold War*. Cambridge: Cambridge University Press.

Levitsky, Steven, and Daniel Ziblatt. 2018. *How Democracies Die*. New York: Crown.

Levy, Jack S. 1983. *War in the Modern Great Power System: 1495–1975*. Lexington: University Press of Kentucky.

Levy, Mickey. 2022. "China Is About to Fall into the Middle-Income Trap." *Wall Street Journal*, October 27.

Lewis, Bernard. 1958. "Some Reflections on the Decline of the Ottoman Empire." *Studia Islamica*, no. 9: 111–27.

Li, Daitian, Tony W. Tong, and Yangao Xiao. 2021. "Is China Emerging as the Global Leader in AI?" *Harvard Business Review*, February 18. https://hbr.org/2021/02/is-china-emerging-as-the-global-leader-in-ai.

Li, David D. 1996. "A Theory of Ambiguous Property Rights in Transition Economies: The Case of the Chinese Non-State Sector." *Journal of Comparative Economics* 23, no. 1: 1–19.

Li, Eric X. 2013. "The Life of the Party: The Post-Democratic Future Begins in China." *Foreign Affairs* 92, no. 1. Accessed April 28, 2022. https://www.foreignaffairshere/articles/china/2012–12–03/life-party.

Li, Shaomin. 2004. "Why Is Property Right Protection Lacking in China? An Institutional Explanation." *California Management Review* 46, no. 3: 100–15.

Li, Yao. 2019. "A Zero-Sum Game? Repression and Protest in China." *Government and Opposition* 54, no. 2: 309–35.

Li, Yuan. 2023. "Why China Didn't Invent Chat GPT." *New York Times*, February 17.

Li, Yuan. 2021. "What China Expects from Businesses: Total Surrender." *New York Times*, July 18.

Li, Daitian, Tony W. Tong, and Yangao Xiao. 2021. "Is China Emerging as the Global Leader in AI?" *Harvard Business Review*, February 18, sec. International Business.

Liao, Kang. 1997. *Pearl S. Buck: A Cultural Bridge Across the Pacific*. Westport, CT: Greenwood Press.

Lieber, Keir A., and Daryl G. Press. 2006. "The End of MAD? The Nuclear Dimension of U.S. Primacy." *International Security* 30, no. 4: 7–44.

Lieber, Keir A., and Daryl G. Press. 2017. "The New Era of Counterforce: Technological Change and the Future of Nuclear Deterrence." *International Security* 41, no. 4: 9–49.

Lieberthal, Kenneth. 1978. *Sino-Soviet Conflict in the 1970s: Its Evolution and Implications for the Strategic Triangle*. Santa Monica: RAND.

Lienert, Ian, and Jitendra R. Modi. 1997. "A Decade of Civil Service Reform in Sub-Saharan Africa." IMF Working Papers 97/179. International Monetary Fund.

Liff, Adam. 2015. "Japan's Defense Policy: Abe the Evolutionary." *Washington Quarterly* 38, no. 2: 79–99.

Lincoln, W. Bruce. 1990. *The Great Reforms: Autocracy, Bureaucracy, and the Politics of Change in Imperial Russia*. DeKalb: Northern Illinois University Press.

Lind, Jennifer. 2018. "Life in China's Asia: What Regional Hegemony Would Look Like." *Foreign Affairs* 97, no. 2: 71–82.

Lind, Jennifer. 2022. "Japan Steps Up: How Asia's Rising Threats Convinced Tokyo to Abandon Its Defense Taboos." *Foreign Affairs*, December 23. Accessed October 8, 2023. https://www.foreignaffairs.com/japan/japan-steps.

Lind, Jennifer. 2023. "Authoritarian Adaptation and Great Power Competition." Henry A. Kissinger Papers. Henry A. Kissinger Center for Global Affairs, John Hopkins University. https://sais.jhu.edu/kissinger/programs-and-projects/kissinger-center-papers/authoritarian-adaptation-great-power-competition.

Lind, Jennifer. 2024. "Back to Bipolarity: How China's Rise Transformed the Balance of Power," *International Security* 49, no. 2: 7–55.

Lind, Jennifer, and Joshua R. Itzkowitz Shifrinson. 2018. "The External Sources of Rising State Strength." Paper written for the Annual Meeting of the International Studies Association, San Francisco, Calif.

Lind, Jennifer, and Michael A. Mastanduno. 2025. "Hard Then, Harder Now: Technology Control Regimes in the Old Cold War and the New." *Texas National Security Review* 8, no. 4.

Lind, Jennifer, and Daryl G. Press. 2020. "Reality Check: American Power in an Age of Constraints." *Foreign Affairs*, April. Accessed September 2, 2020. https://www.foreignaffairs.com/articles/china/2020–02–10/reality-check.

Lind, Jennifer, and Daryl G. Press. 2025. "Strategies of Prioritization: American Foreign Policy After Primacy." *Foreign Affairs* 104, no. 4: 94–107.

Lind, Jennifer, and William C. Wohlforth. 2019. "The Future of the Liberal Order Is Conservative: A Strategy to Save the System." *Foreign Affairs*, April. Accessed September 2, 2020. https://www.foreignaffairs.com/articles/2019–02–12/future-liberal-order-conservative.

Lindblom, Charles E. 1977. *Politics and Markets: The World's Political-Economic Systems*. New York: Basic Books.

Lipset, Seymour Martin. 1959. "Some Social Requisites of Democracy: Economic Development and Political Legitimacy." *American Political Science Review* 53, no. 1: 69–105.

Liu, Feng-chao, Denis Simon, Yu-tao Sun, and Cong Cao. 2011. "China's Innovation Policies: Evolution, Institutional Structure, and Trajectory." *Research Policy* 40, no. 7: 917–31.

Liu, John, and Jin Yu Young. 2023. "What the U.S.-China Chip War Means for a Critical American Ally." *New York Times*, September 27, sec. Business. Accessed October 3, 2023. https://www.nytimes.com/2023/09/27/business/samsung -hynix-south-korea.html.

Liu, Juliana, and Wayne Chang. 2023. "Europe Joins the US in Its Chip War with China." *CNN.com*. Accessed October 14, 2023. https://www.cnn.com/2023/03/09 /tech/china-us-netherlands-chips-curbs-response-hnk-intl/index.html.

Liu, Ling, and Liu, Jun, eds. 1989. *Neo-Authoritarianism: Debate on Theories of Reform*. Beijing: Economic Institution Press.

Liu, Xielin, Sylvia Schwaag Serger, Ulrike Tagscherer, and Amber Y. Chang. 2017. "Beyond Catch-Up—Can a New Innovation Policy Help China Overcome the Middle Income Trap?" *Science and Public Policy* 44, no. 5: 656–69.

Lohmann, Ingrid, and Christine Mayer. 2007. "Educating the Citizen: Two Case Studies on Inclusion and Exclusion in Prussia in the Early Nineteenth Century." *Paedagogica Historica* 43, no. 1: 7–27.

Lorch, Jasmin, and Bettina Bunk. 2017. "Using Civil Society as an Authoritarian Legitimation Strategy: Algeria and Mozambique in Comparative Perspective." *Democratization* 24, no. 6: 987–1005.

Lorentzen, Peter. 2013. "Regularizing Rioting: Permitting Public Protest in an Authoritarian Regime." *Quarterly Journal of Political Science* 8, no. 2: 127–58.

Lorentzen, Peter. 2014. "China's Strategic Censorship." *American Journal of Political Science* 58, no. 2: 402–14.

Loudiy, Fadoua. 2014. *Transitional Justice and Human Rights in Morocco: Negotiating the Years of Lead*. London: Routledge.

Lu, Xiaobo. 2000. "Booty Socialism, Bureau-Preneurs, and the State in Transition: Organizational Corruption in China." *Comparative Politics* 32, no. 3: 273–94.

Lu, Yingdan, and Jennifer Pan. 2021. "Capturing Clicks: How the Chinese Government Uses Clickbait to Compete for Visibility." *Political Communication* 38, no. 1–2: 23–54.

Lumsdaine, David H. 1993. *Moral Vision in International Politics: The Foreign Aid Regime 1949–1989*. Princeton, NJ: Princeton University Press.

Luong, Ngor. 2024. "Testimony Before the US-China Economic and Security Review Commission on 'Current and Emerging Technologies in US-China Economic and National Security Competition." February 1. https://www.uscc .gov/sites/default/files/2024-02/Ngor_Luong_Testimony.pdf.

Lyall, Jason. 2020. *Divided Armies: Inequality and Battlefield Performance in Modern War*. Princeton, NJ: Princeton University Press.

Ma, Jian. 2013. "China's Barbaric One-Child Policy." *Guardian*, May 6.

Ma, Jun. 2017. *The Economics of Air Pollution in China*. New York: Columbia University Press.

Machiavelli, Niccolo. 1992. *The Prince*. New York: Dover.

Mackie, Gerry. 1996. "Ending Footbinding and Infibulation: A Convention Account." *American Sociological Review* 61, no. 6: 999–1017.

MacKinnon, Rebecca. 2011. "China's 'Networked Authoritarianism." *Journal of Democracy* 22, no. 2: 32–46. https://doi.org/10.1353/jod.2011.0033.

Madsen, Elizabeth Leahy. 2012. "Age Structure and Development through a Policy Lens." In *Political Demography: How Population Changes Are Reshaping International Security and National Politics*, edited by Jack A. Goldstone, Eric P. Kaufmann, and Monica Duffy Toft. Boulder, CO: Oxford University Press.

Magnus, George. 2018. *Red Flags: Why Xi's China Is in Jeopardy*. New Haven, CT: Yale University Press.

Mann, Laura, and Marie Berry. 2016. "Understanding the Political Motivations That Shape Rwanda's Emergent Developmental State." *New Political Economy* 21, no. 1: 119–44.

Manyika, James, and Michael Spence. 2023. "The Coming AI Economic Revolution." *Foreign Affairs*, October 24. Accessed November 2, 2023. https://www.foreignaffairs.com/world/coming-ai-economic-revolution.

Mao, Zedong. 1977. *A Critique of Soviet Economy*. Translated by Moss Roberts. New York: Monthly Review Press.

Marinov, Nikolay, and Hein Goemans. 2014. "Coups and Democracy." *British Journal of Political Science* 44, no. 4: 799–825. https://doi.org/10.1017/S0007123413000264.

Martin, Jamie. 2022. *The Meddlers: Sovereignty, Empire, and the Birth of Global Economic Governance*. Cambridge, MA: Harvard University Press. http://www.degruyter.com/document/doi/10.4159/9780674275768/html.

Mastro, Oriana Skylar. 2021. "The Taiwan Temptation: Why China Might Resort to Force." *Foreign Affairs* 100, no. 4: 58–67.

Mastro, Oriana Skylar. 2024. *Upstart: How China Became a Great Power*. Oxford: Oxford University Press.

Matfess, Hilary. 2015. "Rwanda and Ethiopia: Developmental Authoritarianism and the New Politics of African Strong Men." *African Studies Review* 58, no. 2: 181–204. https://doi.org/10.1017/asr.2015.43.

Mati, Armine, and Sidra Rehman. 2022. "Saudi Arabia to Grow at Fastest Pace in a Decade." International Monetary Fund. https://www.imf.org/en/News/Articles/2022/08/09/CF-Saudi-Arabia-to-grow-at-fastest-pace.

Matthews, Mervyn. 1993. *The Passport Society: Controlling Movement in Russia and the USSR*. Boulder, CO: Westview Press.

Mattis, Peter. 2011. "China's Adaptive Approach to the Information Counter Revolution." *Jamestown Foundation China Brief* 11, no. 10: 5–8.

Maynard, Andrew D. 2015. "Navigating the Fourth Industrial Revolution." *Nature Nanotechnology* 10, no. 12: 1005–6.

Mazarr, Michael J., Timothy R. Heath, and Astrid Stuth Cevallos. 2018. *China and the International Order*. Santa Monica, CA: RAND.

McGuire, James W. 2010. *Wealth, Health, and Democracy in East Asia and Latin America*. Cambridge: Cambridge University Press.

McMillan, John, and Paolo Zoido. 2004. "How to Subvert Democracy: Montesinos in Peru." *Journal of Economic Perspectives* 18, no. 4: 69–92.

Mearsheimer, John J. 2014. *The Tragedy of Great Power Politics*. 2nd ed. New York: W. W. Norton.

Mearsheimer, John J. 2021. "The Inevitable Rivalry: America, China, and the Tragedy of Great-Power Politics." *Foreign Affairs* 100, no. 6: 48–59.

Meester, Jos. 2021."'Designed in Ethiopia' and 'Made in China.'" Clingendael. https://www.clingendael.org/publication/designed-ethiopia-and-made-china.

Meng, Anne. 2020. *Constraining Dictatorship: From Personalized Rule to Institutionalized Regimes*. Cambridge: Cambridge University Press.

Michaelsen, Marcus. 2018. "Exit and Voice in a Digital Age: Iran's Exiled Activists and the Authoritarian State." *Globalizations* 15, no. 2: 248–64.

Milesi-Ferretti, G. M., R. Perotti, and M. Rostagno. 2002. "Electoral Systems and Public Spending." *Quarterly Journal of Economics* 117, no. 2: 609–57.

Milgrom, Paul and John Roberts. 1992. *Economics, Organization, and Management*. New York: Prentice-Hall.

Miller, Chris. 2016. *The Struggle to Save the Soviet Economy: Mikhail Gorbachev and the Collapse of the USSR*. Chapel Hill: University of North Carolina Press.

Miller, Chris. 2022a. *Chip War: The Fight for the World's Most Critical Technology*. New York: Scribner.

Miller, Chris. 2022b. "Rewire: Semiconductors and U.S. Industrial Policy." Center for a New American Security, September 19. https://www.cnas.org/publications/reports/rewire-semiconductors-and-u-s-industrial-policy.

Milner, Helen V. and Sondre Ulvund Solstad. 2021. "Technological Change and the International System." *World Politics* 73, no. 3: 545–89. https://doi.org/10.1017/S0043887121000010.

Minzner, Carl. 2018. *End of an Era: How China's Authoritarian Revival Is Undermining Its Rise*. Oxford: Oxford University Press.

Mitcham, Samuel W., Jr. 2008. *The Rise of the Wehrmacht: The German Armed Forces and World War II*. Westport, CT: Praeger Security International.

Mitter, Rana. 2003. "Old Ghosts, New Memories: China's Changing War History in the Era of Post-Mao Politics." *Journal of Contemporary History* 38, no. 1: 117–31.

Mkandawire, Thandika. 2015. "Neopatrimonialism and the Political Economy of Economic Performance in Africa: Critical Reflections." *World Politics* 67, no. 3: 563–612.

Moi, Kho Ee. 2015. "Economic Pragmatism and the "Schooling" of Girls in Singapore." *HSSE Online* 4, no. 2: 62–77. https://hsseonline.nie.edu.sg/economic-pragmatism-and-the-schooling-of-girls-in-singapore/.

Mok, K. H. 2007. "Globalisation, New Education Governance and State Capacity in East Asia." *Globalisation, Societies and Education* 5, no. 1: 1–21.

Mokyr, Joel. 1992. *The Lever of Riches: Technological Creativity and Economic Progress*. Oxford: Oxford University Press.

Mokyr, Joel. 2009. "Intellectual Property Rights, the Industrial Revolution, and the Beginnings of Modern Economic Growth." *The American Economic Review* 99, no. 2: 349–55.

Mokyr, Joel. 2016. *A Culture of Growth: The Origins of the Modern Economy*. Princeton, NJ: Princeton University Press.

Monga, Célestin, and Justin Yifu Lin. 2015. *The Oxford Handbook of Africa and Economics*. Vol. 2, *Policies and Practices*. Oxford: Oxford University Press.

Monteiro, Nuno P. 2014. *Theory of Unipolar Politics*. Cambridge: Cambridge University Press.

Montgomery, Evan Braden. 2014. "Contested Primacy in the Western Pacific: China's Rise and the Future of U.S. Power Projection." *International Security* 38, no. 4: 115–49.

Montinola, Gabriella, Yingyi Qian, and Barry R. Weingast. 1995. "Federalism, Chinese Style: The Political Basis for Economic Success in China." *World Politics* 48, no. 1: 50–81.

Moon, Chung-in. 1994. "Changing Patterns of Business-Government Relations in South Korea." In *Business and Government in Industrialising Asia*, edited by Andrew J MacIntyre. St. Leonards, Australia: Allen & Unwin.

Moon, Chungshik. 2019. "Political Institutions and FDI Inflows in Autocratic Countries." *Democratization* 26, no. 7: 1256–77.

Morgan, Forrest E., Benjamin Boudreaux, Andrew J. Lohn, Mark Ashby, Christian Curriden, Kelly Klima, and Derek Grossman. 2020. *Military Applications of Artificial Intelligence: Ethical Concerns in an Uncertain World*. Santa Monica, CA: RAND. https://www.rand.org/pubs/research_reports/RR3139-1.html.

Morgenbesser, Lee. 2020. "The Menu of Autocratic Innovation." *Democratization* 27, no. 6: 1053–72.

Moss, Trefor. 2021. "NIO, the Chinese Electric-Vehicle Startup, Unveils New ET7 Sedan." *Wall Street Journal*, January 9.

Mosse, W. E. 1966. *Alexander II and the Modernization of Russia*. New York: Macmillan.

Mounk, Yascha. 2018a. "America Is Not a Democracy." *Atlantic*, March. https://www.theatlantic.com/magazine/archive/2018/03/america-is-not-a-democracy/550931.

Mounk, Yascha. 2018b. *The People vs. Democracy: Why Our Freedom Is in Danger and How to Save It*. Cambridge, MA: Harvard University Press.

Moustafa, Tamir. 2007. *The Struggle for Constitutional Power: Law, Politics, and Economic Development in Egypt*. Cambridge: Cambridge University Press.

Moyo, Dambisa. 2018. *Edge of Chaos: Why Democracy Is Failing to Deliver Economic Growth—and How to Fix It*. New York: Basic Books.

Mulligan, Casey B., Ricard Gil, and Xavier Sala-i-Martin. 2004. "Do Democracies Have Different Public Policies than Nondemocracies?" *Journal of Economic Perspectives* 18, no. 1: 51–74.

Murray, Michelle. 2018. *The Struggle for Recognition in International Relations: Status, Revisionism, and Rising Powers*. Oxford: Oxford University Press.

Murray, Williamson A. 2005. "The Industrialization of War 1815–1871." In *The Cambridge History of Warfare*, Vol. 2, edited by Aaron Sheehan-Dean. Cambridge: Cambridge University Press.

Murray, Williamson and Allan R. Millett, eds. 1998. *Military Innovation in the Interwar Period*. Cambridge: Cambridge University Press.

Nanes, Stefanie Eileen. 2008. "Fighting Honor Crimes: Evidence of Civil Society in Jordan." In *Deconstructing Sexuality in the Middle East : Challenges and Discourses*, edited by Pinar Ilkkaracan. London: Routledge.

Nathan, Andrew J. 2003. "China's Changing of the Guard: Authoritarian Resilience." *Journal of Democracy* 14, no. 1: 6–17.

National Science Board, National Science Foundation. 2024. Research and Development: U.S. Trends and International Comparisons. *Science and Engineering Indicators*. NSB-2024-6. Alexandria, VA. https://ncses.nsf.gov/pubs/nsb20246/.

National Science Board, National Science Foundation. 2023. Publications Output: U.S. Trends and International Comparisons. *Science and Engineering Indicators 2024*. NSB-2023-33. Alexandria, VA. https://ncses.nsf.gov/pubs/nsb202333/.

Naughton, Barry. 1995. *Growing Out of the Plan: Chinese Economic Reform, 1978–1993*. Cambridge: Cambridge University Press.

Naughton, Barry. 2008. "A Political Economy of China's Economic Transition." In *China's Great Economic Transformation*, edited by Loren Brandt and Thomas G. Rawski. Cambridge: Cambridge University Press.

Nazarbayev, Nursultan. 2008. *The Kazakhstan Way*. Translated by Jan Butler. London: Stacey International.

Nelson, Richard R., and Howard Pack. 1999. "The Asian Miracle and Modern Growth Theory." *Economic Journal* 109, no. 457: 416–36.

"A New Course for the Republic." Government of Kazakhstan. https:// kazakhstan2050.com/.

Neve, Freddie. 2021. "Economic and Social Reform in the UAE." Asia House, June 8. https://asiahouse.org/news-and-views/economic-and-social-reform-in-the -uae-a-bid-to-boost-growth-and-investment/.

Ngai, Joseph Luc, John Qu, and Nicole Zhou. 2016. "What's Next for China's Booming Fintech Sector?" McKinsey & Company, July 2016. https://www .mckinsey.com/~/media/mckinsey/industries/financial%20services/our%20 insights/whats%20next%20for%20chinas%20booming%20fintech%20sector /whats-next-for-chinas-booming-fintech-sector.pdf.

Niida, Noboru. 1964. "Land Reform and New Marriage Law in China." *Developing Economies* 2, no. 1: 3–15. https://doi.org/10.1111/j.1746-1049.1964.tb00667.x.

Nixon, Richard M. 1967. "Asia after Viet Nam." *Foreign Affairs* 46, no. 1: 111–25.

North, Douglass C. 1990. *Institutions, Institutional Change and Economic Performance*. Cambridge: Cambridge University Press.

North, Douglass C., and Robert Paul Thomas. 1973. *The Rise of the Western World: A New Economic History*. Cambridge: Cambridge University Press.

North, Douglass C., John Joseph Wallis, and Barry R. Weingast. 2012. *Violence and Social Orders: A Conceptual Framework for Interpreting Recorded Human History*. Cambridge: Cambridge University Press.

North, Douglass C., and Barry R. Weingast. 1989. "Constitutions and Commitment: The Evolution of Institutions Governing Public Choice in Seventeenth-Century England." *Journal of Economic History* 49, no. 4: 803–32.

Nye, Joseph S. 2004. *Soft Power: The Means to Success in World Politics*. New York: PublicAffairs.

O'Brien, Kevin, and Yanhua Deng. 2017. "Preventing Protest One Person at a Time: Psychological Coercion and Relational Repression in China." *China Review* 17, no. 2: 179–201.

O'Donnell, Guillermo, and Philippe Schmitter. 1986. *Transitions from Authoritarian Rule: Tentative Conclusions About Uncertain Democracies*. Baltimore, MD: John Hopkins University Press.

O'Mara, Margaret. 2015. *Cities of Knowledge: Cold War Science and the Search for the Next Silicon Valley*. Princeton, NJ: Princeton University Press. https://www .degruyter.com/document/doi/10.1515/9781400866885/html.

O'Rourke, Lindsey A. 2018. *Covert Regime Change: America's Secret Cold War*. Ithaca, NY: Cornell University Press.

Office of the Secretary of Defense. 2022. *Military and Security Developments Involving the People's Republic of China*. Annual Report to Congress. Washington, DC: US Department of Defense. https://www.defense.gov/Spotlights/2022-China -Military-Power-Report/.

Ohlin, Jens David, and Duncan B. Hollis. 2021. *Defending Democracies: Combating Foreign Election Interference in a Digital Age*. Oxford: Oxford University Press.

Olar, Roman-Gabriel. 2019. "Do They Know Something We Don't? Diffusion of Repression in Authoritarian Regimes." *Journal of Peace Research* 56, no. 5: 667–81.

Olson, Mancur. 1982. *The Rise and Decline of Nations: Economic Growth, Stagflation, and Social Rigidities*. New Haven, CT: Yale University Press.

Olson, Mancur. 1993. "Dictatorship, Democracy, and Development." *American Political Science Review* 87, no. 3: 567–76.

Olson, Mancur. 2000. *Power and Prosperity: Outgrowing Communist and Capitalist Dictatorships*. New York: Basic Books.

Olson, Mancur. 1965. *The Logic of Collective Action: Public Goods and the Theory of Groups*. Cambridge, MA: Harvard University Press.

Ong, Lynette H. 2022. *Outsourcing Repression: Everyday State Power in Contemporary China*. Oxford: Oxford University Press.

Ooi, Kee Beng. 2010. *In Lieu of Ideology: An Intellectual Biography of Goh Keng Swee*. Cambridge: Cambridge University Press.

Orf, Darren. 2025. "This 'Artificial Sun' Just Smashed Its Own Nuclear Fusion Record." *Popular Mechanics*, January 24, 2025. https://www.popularmechanics.com/science/green-tech/a63512763/nuclear-fusion-east-china/?utm_source=701687&utm_medium=email.

Organski, A. F. K., and Jacek Kugler. 1980. *The War Ledger*. Chicago: University of Chicago Press.

Ottaway, David. 2021. "Saudi Crown Prince Lambasts His Kingdom's Wahhabi Establishment." *Viewpoints Series*, May 6. Wilson Center. https://www.wilsoncenter.org/article/saudi-crown-prince-lambasts-his-kingdoms-wahhabi-establishment.

Ottaway, Marina, and Michele Dunne. 2007. "Incumbent Regimes and the 'King's Dilemma' in the Arab World: Promise and Threat of Managed Reform." Carnegie Papers No. 88. Middle East Program, Carnegie Endowment for International Peace.

Oud, Malin. 2020. "Harmonic Convergence: China and the Right to Development." In *An Emerging China-Centric Order: China's Vision for a New World Order in Practice*, edited by Nadège Rolland. Seattle, WA: National Bureau of Asian Research.

Overholt, William H. 2018. *China's Crisis of Success*. Cambridge: Cambridge University Press.

Pace, Eric. 1984. "Ahmed Sekou Toure, A Radical Hero." *New York Times*, March 28.

Pack, Howard, and Kamal Saggi. 2006. "Is There a Case for Industrial Policy? A Critical Survey." *World Bank Research Observer* 21, no. 2: 267–97.

Packard, George. 2010. *Edwin O. Reischauer and the American Discovery of Japan*. New York: Columbia University Press.

Paglayan, Agustina S. 2021. "The Non-Democratic Roots of Mass Education: Evidence from 200 Years." *American Political Science Review* 115, no. 1: 179–98.

Palen, John J. 1986. "Fertility and Eugenics: Singapore's Population Policies." *Population Research and Policy Review* 5, no. 1: 3–14.

Pamuk, Şevket, and Jeffrey G. Williamson. 2011. "Ottoman De-Industrialization, 1800–1913: Assessing the Magnitude, Impact, and Response." *Economic History Review* 64, no. S1: 159–84.

Paul, T. V., Deborah Welch Larson, and William C. Wohlforth. 2014. *Status in World Politics.* Cambridge: Cambridge University Press.

Pavitt, Keith. 1984. "Sectoral Patterns of Technical Change: Towards a Taxonomy and a Theory." *Research Policy* 13, no. 6: 343–73.

Pearson, Margaret M., Meg Rithmire, and Kellee S. Tsai. 2022. "China's Party-State Capitalism and International Backlash: From Interdependence to Insecurity." *International Security* 47, no. 2: 135–76.

Peattie, Mark. 2013. *Sunburst: The Rise of Japanese Naval Air Power, 1909–1941.* Annapolis, MD: Naval Institute Press.

Peck, Jennifer. 2024. "Working Women are Changing Saudi Arabia." *Foreign Affairs,* June 19.

Pei, Minxin. 2000. "Rights and Resistance: The Changing Contexts of the Dissident Movement." In *Chinese Society: Change, Conflict and Resistance.* Vol. 2, edited by Elizabeth J. Perry and Mark Selden. London: Routledge.

Pei, Minxin. 2006. *China's Trapped Transition.* Cambridge, MA: Harvard University Press.

Pepinsky, Thomas. 2020. "Authoritarian Innovations: Theoretical Foundations and Practical Implications." *Democratization* 27, no. 6: 1092–101.

Pereira, N. G. O. 1983. *Tsar-Liberator: Alexander II of Russia 1818–1881.* Newtonville, MA: Oriental Research Partners.

Perry, William J. 2004. "Military Technology: An Historical Perspective." *Technology in Society* 26, no. 2–3: 235–43.

"Peru Strips Controversial TV Mogul of Citizenship." 1997. *Wall Street Journal,* July 14. https://www.wsj.com/articles/SB868855349478394500.

Peterson, Glen. 1994. "State Literacy Ideologies and the Transformation of Rural China." *Australian Journal of Chinese Affairs,* no. 32: 95–120.

Petracca, Mark P., and Mong Xiong. 1990. "The Concept of Chinese Neo-Authoritarianism: An Exploration and Democratic Critique." *Asian Survey* 30, no. 11: 1099–117.

Pettis, Michael. 2011. "China's Troubled Transition to a More Balanced Growth Model." Policy Paper. New America Foundation, March 2011. https://www.newamerica.org/economic-growth/policy-papers/chinas-troubled-transition-to-a-more-balanced-growth-model/.

Pickard, Victor. 2019. *Democracy Without Journalism?: Confronting the Misinformation Society.* Oxford: Oxford University Press.

Poats, Rutherford M. 1985. *Twenty-Five Years of Development Cooperation.* OECD.

Policzer, Pablo. 2009. *The Rise and Fall of Repression in Chile.* South Bend, IN: University of Notre Dame Press.

Pollpeter, Kevin L., Michael S. Chase, and Eric Heginbotham. 2017. "The Creation of the PLA Strategic Support Force and Its Implications for Chinese Military Space Operations." Washington, DC: RAND. Accessed November 4, 2022. https://www.rand.org/pubs/research_reports/RR2058.html.

Polyakova, Alina, and Chris Meserole. 2019. "Exporting Digital Authoritarianism." Washington, DC: Brookings Institution.

Pomeranz, Kenneth. 2000. *The Great Divergence: China, Europe, and the Making of the Modern World Economy*. Princeton, NJ: Princeton University Press.

Pop, Valentina, Sha Hua, and Daniel Michaels. 2021. "From Lightbulbs to 5G, China Battles West for Control of Vital Technology Standards." *Wall Street Journal*, February 8. Accessed August 12, 2021. https://www.wsj.com/articles/from-lightbulbs-to-5g-china-battles-west-for-control-of-vital-technology-standards-11612722698.

Popper, Karl R. 2020. *The Open Society and Its Enemies*. Princeton, NJ:Princeton University Press.

Porter, Michael E. 1990. *Competitive Advantage of Nations: Creating and Sustaining Superior Performance*. New York: Simon and Schuster.

Posen, Barry R. 2011. "From Unipolarity to Multipolarity: Transition in Sight?" In *International Relations Theory and the Consequences of Unipolarity*, edited by G. John Ikenberry, Michael Mastanduno, and William C. Wohlforth. Cambridge: Cambridge University Press.

Pritchett, Lant, and Lawrence H. Summers. 2014. "Asiaphoria Meets Regression to the Mean." Working Paper No. 20573. National Bureau of Economic Research. Accessed March 28, 2022. https://www.nber.org/papers/w20573.Przeworski, Adam. 2004. "Geography vs. Institutions Revisited: Were Fortunes Reversed?" Working Paper. New York University. https://as.nyu.edu/content/dam/nyu-as/faculty/documents/reversal.pdf.

Przeworski, Adam, Michael E. Alvarez, and Jose Antonio Cheibub. 2000. *Democracy and Development: Political Institutions and Well-Being in the World, 1950–1990*. New York: Cambridge University Press.

Przeworski, Adam, and Fernando Limongi. 1993. "Political Regimes and Economic Growth." *Journal of Economic Perspectives* 7, no. 3: 51–69.

Pyle, Jean L. 1997. "Women, the Family, and Economic Restructuring: The Singapore Model?" *Review of Social Economy* 55, no. 2: 215–23.

Quah, Jon S. T. 1999. "Corruption in Asian Countries: Can It Be Minimized?" *Public Administration Review* 59, no. 6: 483–94. https://doi.org/10.2307/3110297.

Quinlivan, James T. 1999. "Coup-Proofing: Its Practice and Consequences in the Middle East." *International Security* 24, no. 2: 131–65.

"Raffles Statue." Visit Singapore. https://www.visitsingapore.com.cn/content/mobile/en/see-do-singapore/history/memorials/sir-raffles-statue-landing-site/.

Rajah, Jothie. 2012. *Authoritarian Rule of Law: Legislation, Discourse and Legitimacy in Singapore*. Cambridge: Cambridge University Press.

Rajan, R., and L. Zingales. 1998. "Financial Dependence and Growth." *American Economic Review* 88, no. 3: 559–86.

Rajaratnam, S. 1972. "'Singapore: The Global City': Speech given to the Singapore Press Club on 6 February 1972." In *S. Rajaratnam: The Prophetic and the Political*, edited by Chan Heng Chee and O. Hag. Singapore: Graham Brash.

Ramzy, Austin, and Chris Buckley. 2019. "'Absolutely No Mercy': Leaked Files Expose How China Organized Mass Detentions of Muslims." *New York Times*, November 16, sec. World. Accessed December 30, 2022. https://www.nytimes.com/interactive/2019/11/16/world/asia/china-xinjiang-documents.html.

Rana, Pradumna Bikram, and Chiya-yi Lee. 2015. "Economic Legacy of Lee Kuan Yew: Lessons for Aspiring Countries." *RSIS Commentary*, March 31. https://www.rsis.edu.sg/rsis-publication/cms/co15073-economic-legacy-of-lee-kuan-yew-lessons-for-aspiring-countries/#.YySLaezMJz8.

Rasler, Karen A., and William R. Thompson. 1994. *The Great Powers and Global Struggle, 1490–1990*. Lexington: University Press of Kentucky.

Ravina, Mark. 2017. *To Stand with the Nations of the World: Japan's Meiji Restoration in World History*. Oxford: Oxford University Press.

Reif, L. Raphael. 2021. "How to Build on Vannevar Bush's 'Wild Garden' to Cultivate Solutions to Human Needs." *Issues in Science and Technology*, July 12, 2021. https://issues.org/vannevar-bush-wild-garden-science-policy-reif/.

Reiter, Dan. 2020. "Avoiding the Coup-Proofing Dilemma: Consolidating Political Control While Maximizing Military Power." *Foreign Policy Analysis* 16, no. 3: 312–31.

Repucci, Sarah, and Amy Slipowitz. 2022. *Freedom in the World 2022: The Global Expansion of Authoritarian Rule*. Washington, DC: Freedom House. https://freedomhouse.org/report/freedom-world/2022/global-expansion-authoritarian-rule.

Richter, Linda K. 1989. "About Face: The Political Evolution of Chinese Tourism Policy." *The Politics of Tourism in Asia*. University of Hawai'i Press. Accessed November 10, 2022. https://www.jstor.org/stable/j.ctv9zcjr9.6.

Roberts, Margaret E. 2018. *Censored: Distraction and Diversion Inside China's Great Firewall*. Princeton, NJ: Princeton University Press.

Robertson, Jordan. 2023. "China's Semiconductor Ambitions Fuel European Brain Drain." Bloomberg.com, July 19. Accessed October 14, 2023. https://www.bloomberg.com/news/newsletters/2023–07–19/china-s-semiconductor-ambitions-fuel-european-brain-drain.

Rock, Michael T. 2017. *Dictators, Democrats, and Development in Southeast Asia: Implications for the Rest*. Oxford University Press.

Rodan, Garry. 1996. "Elections without Representation: The Singapore Experience under the PAP." In *The Politics of Elections in Southeast Asia*, edited by R. H. Taylor. Cambridge: Cambridge University Press.

Rodriguez, Sarah Mellors. 2023. *Reproductive Realities in Modern China: Birth Control and Abortion, 1911–2021*. Cambridge: Cambridge University Press.

Rodrik, Dani. 1992. "The Rush to Free Trade in the Developing World: Why So Late? Why Now? Will It Last?" Working Paper No. 3947. National Bureau of Economic Research. Accessed September 12, 2022. https://www.nber.org/papers/w3947.

Rodrik, Dani. 2007. *One Economics, Many Recipes: Globalization, Institutions, and Economic Growth*. Princeton, NJ: Princeton University Press.

Rodrik, Dani, Arvind Subramanian, and Francesco Trebbi. 2002. "Institutions Rule: The Primacy of Institutions over Geography and Integration in Economic Development." Working Paper No. 9305. National Bureau of Economic Research. https://www.nber.org/papers/w9305.

Rogoff, Kenneth. 1985. "The Optimal Degree of Commitment to an Intermediate Monetary Target." *Quarterly Journal of Economics* 100, no. 4: 1169–89.

Rolland, Nadège. 2020. "An Emerging China-Centric Order: China's Vision for a New World Order in Practice." Seattle, WA: National Bureau of Asian Research.

Rosen, Stanley. 1984. "New Directions in Secondary Education." In *Contemporary Chinese Education*, edited by Ruth Hayhoe. Croom Helm.

Rosenfeld, Bryn. 2017. "Reevaluating the Middle-Class Protest Paradigm: A Case-Control Study of Democratic Protest Coalitions in Russia." *American Political Science Review* 111, no. 4: 637–52.

Rosenfeld, Bryn. 2020. *The Autocratic Middle Class.* Princeton, NJ: Princeton University Press.

Ross, Michael. 2006. "Is Democracy Good for the Poor?" *American Journal of Political Science* 50, no. 4: 860–74.

Rostow, Walt W. 1960. *The Stages of Economic Growth: A Non-Communist Manifesto.* Cambridge: Cambridge University Press.

Rotman, David. 2023. "A New US Innovation Narrative." *Technology Review* 126, no. 1: 66–69.

Rozelle, Scott, and Natalie Hell. 2020. *Invisible China: How the Urban-Rural Divide Threatens China's Rise.* Chicago: University of Chicago Press.

Rudd, Kevin. 2022. "The Return of Red China." *Foreign Affairs.* November 9. https://www.foreignaffairs.com/china/return-red-china.

Rühlig, Tim. 2022. "Chinese Influence through Technical Standardization Power." *Journal of Contemporary China* 32, no 139: 54–72. https://doi.org/10.1080/10670564.2022.2052439.

Rühlig, Tim. 2023. "The Sources of China's Innovativeness. Why China's 'Unstoppable' Innovation Powerhouse Might Falter." *DGAP Analysis*, no. 5. Accessed August 13, 2024. https://dgap.org/en/research/publications/sources-chinas-innovativeness.

Ruwich, John. 2022. "The Significance of Beijing Hosting Both the Summer and Winter Olympics." NPR.org, January 26, 2022.

Ryan, Curtis R. 2022. "Jordan: A Perpetually Liberalising Autocracy." In *New Authoritarian Practices in the Middle East and North Africa*, edited by Francesco Cavatorta, Merouan Mekouar, and Ozgun Topak. Edinburgh: Edinburgh University Press.

Ryugen, Hideaki, and Hiroyuki Akiyama. 2020. "China Leads the Way in Standards Setting in 5G and Beyond." *Financial Times*, August 4.

Sacks, David. 2022. "What Is China Learning from Russia's War in Ukraine?" *Foreign Affairs*, May 16. Accessed July 19, 2022. https://www.foreignaffairs.com/articles/china/2022–05–16/what-china-learning-russias-war-ukraine.

Saich, Tony. 2000. "Negotiating the State: The Development of Social Organizations in China." *China Quarterly*, no. 161: 124–41.

Saich, Tony. 2021. "Tony Saich on the Party and Private Business: Lessons from History." Ash Center for Democratic Governance and Innovation. Harvard Kennedy School. https://ash.harvard.edu/party-and-private-business-lessons-history.

Saiegh, Sebastian M. 2005. "Do Countries Have a 'Democratic Advantage'?: Political Institutions, Multilateral Agencies, and Sovereign Borrowing." *Comparative Political Studies* 38, no. 4: 366–87.

Samila, S., and O. Sorenson. 2011. "Venture Capital, Entrepreneurship and Economic Growth." *Review of Economics and Statistics* 93, no. 1: 338–49.

Sanborn, Howard, and Clayton L. Thyne. 2014. "Learning Democracy: Education and the Fall of Authoritarian Regimes." *British Journal of Political Science* 44, no. 4: 773–97.

Sautman, Barry. 1992. "Sirens of the Strongman: Neo-Authoritarianism in Recent Chinese Political Theory." *China Quarterly*, no. 129: 72–102.

Saw, Swee-Hock. 1980. "The Development of Population Control in Singapore." *Contemporary Southeast Asia* 1, no. 4: 348–66.

Saw, Swee-Hock. 2012. *The Population of Singapore*. 3rd ed. Singapore: Institute of Southeast Asian Studies.

Saywell, William G. 1980. "Education in China Since Mao." *Canadian Journal of Higher Education* 10, no. 1: 1–27.

Schake, Kori. 2017. *Safe Passage: The Transition from British to American Hegemony*. Cambridge, MA: Harvard University Press.

Schamis, Hector E. 1999. "Distributional Coalitions and the Politics of Economic Reform in Latin America." *World Politics* 51, no. 2: 236–68.

Schell, Orville, and John Delury. 2014. *Wealth and Power: China's Long March to the Twenty-First Century*. New York: Random House.

Schiek, Sebastian. 2018. "Uzbekistan's Transformation from an 'Old' to an 'Upgraded' Autocracy." *L'Europe en Formation* 385, no. 1: 87–103.

Schiek, Sebastian. 2022. "The Politics of Stability in Kazakhstan: Depoliticising Participation Through Consultative Ideology?" *Europe-Asia Studies* 74, no. 2: 266–87. https://doi.org/10.1080/09668136.2022.2034746.

Schleunes, Karl A. 1979. "Enlightenment, Reform, Reaction: The Schooling Revolution in Prussia." *Central European History* 12, no. 4: 315–42.

Schmid, Jon, and Jonathan Huang. 2017. "State Adoption of Transformative Technology: Early Railroad Adoption in China and Japan." *International Studies Quarterly* 61, no. 3: 570–83.

Schmidt, Eric, and Yll Bajraktari. 2022. "America Could Lose the Tech Contest with China: How Washington Can Craft a New Strategy." *Foreign Affairs*, September 8. https://www.foreignaffairs.com/united-states/america-losing-its-tech-contest-china.

Schneider, Ben Ross. 1999. "The Desarrollista State in Brazil and Mexico." In *The Developmental State*, edited by Meredith Woo-Cumings. Ithaca, NY: Cornell University Press.

Schueth, Sam. 2015. "Winning the Rankings Game: The Republic of Georgia, USAID, and the Doing Business Project." In *Ranking the World*, edited by Alexander Cooley and Jack Snyder. Cambridge: Cambridge University Press.

Schultze, Charles L. 1983. "Industrial Policy: A Dissent." *Brookings Review* 2, no. 1: 3–12.

Schumpeter, Joseph A. 1911. *A Theory of Economic Development*. Cambridge, MA: Harvard University Press.

Schumpeter, Joseph A. 1934. *The Theory of Economic Development. An Inquiry into Profits, Capital, Credit, Interest, and the Business Cycle*. Cambridge: Harvard University Press.

Schwab, Klaus. 2017. *The Fourth Industrial Revolution*. New York: Crown Business.

Schwarck, Edward. 2018. "Intelligence and Informatization: The Rise of the Ministry of Public Security in Intelligence Work in China." *China Journal* 80: 1–23.

Schweller, Randall L. 1992. "Domestic Structure and Preventive War: Are Democracies More Pacific?" *World Politics* 44, no. 2: 235–69.

Sciubba, Jennifer D. 2022. *8 Billion and Counting: How Sex, Death, and Migration Shape Our World*. New York: W. W. Norton.

"Secretary of State Antony J. Blinken's remarks at the National Security Commission on Artificial Intelligence's (NSCAI) Global Emerging Technology Summit." Washington, DC, July 13, 2021.

Segal, Adam. 2011. *Advantage: How American Innovation Can Overcome the Asian Challenge*. New York: W. W. Norton.

Seow, Adeline, and Hin-Pin Lee. 1994. "From Colony to City State: Changes in Health Needs in Singapore from 1950 to 1990." *Journal of Public Health Medicine* 16, no. 2: 149–58.

Setser, Brad. 2022. "Beijing's Debts Come Due." *Foreign Affairs*, August 30.

Shambaugh, David. 2008. *China's Communist Party: Atrophy and Adaptation*. Washington, DC: Woodrow Wilson Center Press.

Shambaugh, David. 2015. "The Coming Chinese Crackup." *Wall Street Journal*, March 6.

Shambaugh, David. 2016. *China's Future*. London: John Wiley.

Shambaugh, David. 2021. *China's Leaders: From Mao to Now*. Cambridge: Polity Press.

Shambaugh, David. 2022. "Becoming a Ganbu: China's Cadre Training School System." *Journal of Contemporary China* 32, no. 142: 540–58. https://doi.org/10.1080/10670564.2022.2109008.

Sharivkan, Zardenbek, Bagdat S. Rakhmetulina, Erlan A. Naurysbaev, Berik A. Nukiev, and Aiym M. Aimukanova. 2016. "Analysis of the Main Trends in the Development of Civil Society Institutions in the Republic of Kazakhstan." *International Electronic Journal of Mathematics Education* 11 no. 7: 2203–12.

Sharma, Ruchir. 2012. *Breakout Nations: In Pursuit of the Next Economic Miracles*. New York: W. W. Norton.

Sharma, Ruchir. 2016. "The Demographics of Stagnation: Why People Matter for Economic Growth." *Foreign Affairs* 95, no. 2: 18–24.

Shatz, Howard J., and David J. Tarr. 2000. "Exchange Rate Overvaluation and Trade Protection: Lessons from Experience." Policy Research Working Paper No. 2289. World Bank, Washington, DC.

Shearer, D. 2004. "Elements Near and Alien: Passportization, Policing, and Identity in the Stalinist State, 1932–1952." *Journal of Modern History* 76, no. 4: 835–81.

Sheehan, Matt. 2023. "How China Became an Innovation Powerhouse." Carnegie Endowment for International Peace. Accessed March 22, 2023. https://carnegieendowment.org/2023/01/10/how-china-became-innovation-powerhouse-pub-88761.

Shifrinson, Joshua R. Itzkowitz. 2018. *Rising Titans, Falling Giants: How Great Powers Exploit Power Shifts*. Ithaca, NY: Cornell University Press.

Shirk, Susan L. 1993. *The Political Logic of Economic Reform in China*. Berkeley: University of California Press.

Shirk, Susan L. 2010. "Changing Media, Changing China." In *Changing Media, Changing China*, edited by Susan L. Shirk. Oxford: Oxford University Press.

Shirk, Susan L. 2017. "Xi Jinping's U-Turn to Personalist Rule." China Focus, University of California, San Diego, August 30. https://chinafocus.ucsd.edu/2017/08/30/xi-jinpings-u-turn-to-personalistic-rule/.

Shirk, Susan L. 2018. "China in Xi 'New Era': The Return to Personalistic Rule." *Journal of Democracy* 29, no. 2: 22–36.

Shirk, Susan L. 2022a. *Overreach: How China Derailed Its Peaceful Rise*. Oxford: Oxford University Press.

Shirk, Susan L. 2022b. "Xi Jinping Has Fallen into the Dictator Trap." *New York Times*, October 14, sec. Opinion. Accessed November 5, 2022. https://www.nytimes.com/2022/10/14/opinion/international-world/china-politics-xi.html.

Sidel, Victor W. 1972. "The Barefoot Doctors of the People's Republic of China." *New England Journal of Medicine* 286, no. 24: 1292–300.

Sievert, Jacqueline. 2018. "The Case for Courts: Resolving Information Problems in Authoritarian Regimes." *Journal of Peace Research* 55, no. 6: 774–86. https://doi.org/10.1177/0022343318770236.

Silverstein, Gordon. 2008. "Singapore: The Exception That Proves Rules Matter." In *Rule by Law: The Politics of Courts in Authoritarian Regimes*, edited by Tom Ginsburg and Tamir Moustafa. Cambridge: Cambridge University Press.

"Singapore." Freedom in the World 2021. Freedom House. https://freedomhouse.org/country/singapore/freedom-world/2021.

"Singapore." Index of Economic Freedom. Heritage Foundation, October 2023. Updated February 2025. https://www.heritage.org/index/country/singapore.

"Singapore Ranking in the Global Innovation Index." 2023. World Intellectual Property Organization. https://www.wipo.int/gii-ranking/en/singapore.

"Singapore: Social Media Companies Forced to Cooperate with Abusive Fake News Law." 2020. Amnesty International, February 19. https://www.amnesty.org/en/latest/news/2020/02/singapore-social-media-abusive-fake-news-law/.

"Singapore's Global Innovation Ranking." 2024. Intellectual Property Office of Singapore. https://www.ipos.gov.sg/resources/singapore-ip-ranking#:~:text=Overall%2C%20Singapore%20is%20ranked%20first,Singapore%20second%20in%20the%20world.

Sinkkonen, Elina. 2021. "Dynamic Dictators: Improving the Research Agenda on Autocratization and Authoritarian Resilience." *Democratization* 28, no. 6: 1172–90.

Siripurapu, Anshu, and Noah Berman. 2023. "Is U.S. Industrial Policy Making a Comeback?" Council on Foreign Relations, September 18. https://www.cfr.org/backgrounder/industrial-policy-making-comeback.

Skinner, Jonathan, and Douglas Staiger. 2007. "Technology Adoption from Hybrid Corn to Beta-Blockers." In *Hard-to-Measure Goods and Services: Essays in Honor of Zvi Griliches*, edited by Ernst R. Berndt and Charles R. Hulten. Chicago: University of Chicago Press.

Smith, Keith. 2006. "Measuring Innovation." In *The Oxford Handbook of Innovation*, edited by Jan Fagerberg, David Mowrey, and Richard R. Nelson. Oxford: Oxford University Press. Accessed October 20, 2022. https://smartech.gatech.edu/handle/1853/43180.

Solomon, Peter H. Jr. 2007. "Courts and Judges in Authoritarian Regimes." *World Politics* 60, no. 1: 122–45.

Soltysinski, S. J. 1969. "New Forms of Protection for Intellectual Property in the Soviet Union and Czechoslovakia." *Modern Law Review* 32, no. 4: 408–19.

Spence, Jonathan D. 1999a. *Mao Zedong: A Life*. New York: Penguin.

Spence, Jonathan D. 1999b. *The Search for Modern China*. 2nd ed. New York: W. W. Norton.

Spufford, Francis. 2012. *Red Plenty*. Minneapolis: Graywolf Press.

"Statistics on Chinese Learners Studying Overseas in 2019." 2020. People's Republic of China. Ministry of Education, December 15. http://en.moe.gov.cn /news/press_releases/202012/t20201224_507474.html.

Stevenson, Alexandra, and Zixu Wang. 2023. "China's Population Falls, Heralding a Demographic Crisis." *New York Times*, January 17.

Stockholm International Peace Research Institute. 2024. *SIPRI Yearbook 2024: Armaments, Disarmament and International Security*. New York: Oxford University Press. https://www.sipri.org/sites/default/files/2024–06/yb24_summary_en _2_1.pdf.

Strange, Susan. 1987. "The Persistent Myth of Lost Hegemony." *International Organization* 41, no. 4: 551–74.

Subotić, Jelena. 2009. *Hijacked Justice: Dealing with the Past in the Balkans*. Ithaca, NY: Cornell University Press.

Subramanian, Arvind. 2011. "The Inevitable Superpower." *Foreign Affairs*, August 19.

Sun, Yan. 1999. "Reform, State, and Corruption: Is Corruption Less Destructive in China than in Russia?" *Comparative Politics* 32, no. 1: 1–20.

Suzuki, Ken. 2018. "China's New 'Xi Jinping Constitution': The Road to Totalitarianism." Nippon.com, November 27. https://www.nippon.com/en /in-depth/a05803/.

Svolik, Milan W. 2012. *The Politics of Authoritarian Rule*. Cambridge: Cambridge University Press.

Swanson, Ana, David McCabe, and Michael Crowley. 2023. "Biden Administration Weighs Further Curbs on Sales of A.I. Chips to China." *New York Times*, June 28. Accessed August 8, 2023. https://www.nytimes.com/2023/06/28/business /economy/biden-administration-ai-chips-china.html.

Szabłowski, Witold. 2020. *How to Feed a Dictator: Saddam Hussein, Idi Amin, Enver Hoxha, Fidel Castro, and Pol Pot Through the Eyes of Their Cooks*. New York: Penguin.

Talmadge, Caitlin. 2015. *The Dictator's Army: Battlefield Effectiveness in Authoritarian Regimes*. Ithaca, NY: Cornell University Press.

Tamanaha, Brian Z. 2004. *On the Rule of Law: History, Politics, Theory*. Cambridge: Cambridge University Press.

Tan, Kenneth Paul. 2012. "The Ideology of Pragmatism: Neo-Liberal Globalisation and Political Authoritarianism in Singapore." *Journal of Contemporary Asia* 42, no. 1: 67–92.

Tanner, Murray Scot.2000. "State Coercion and the Balance of Awe: The 1983–1986 'Stern Blows' Anti-Crime Campaign." *China Journal*, no. 44: 93–125.

Tansey, Oisín. 2016. *The International Politics of Authoritarian Rule*. Oxford: Oxford University Press. Accessed December 10, 2022. https://doi.org/10.1093/acprof: oso/9780199683628.003.0007.

Tapscott, Rebecca. 2021. *Arbitrary States: Social Control and Modern Authoritarianism in Museveni's Uganda*. Oxford: Oxford University Press.

Taylor, Mark Zachary. 2016. *The Politics of Innovation: Why Some Countries Are Better Than Others at Science and Technology*. Oxford: Oxford University Press.

Teets, Jessica C. 2014. *Civil Society Under Authoritarianism: The China Model*. Cambridge: Cambridge University Press.

Tellis, Ashley J., Janice Bially, Christopher Layne, and Melissa McPherson. 2000. *Measuring National Power in the Postindustrial Age.* Santa Monica, CA: RAND. Accessed April 10, 2022. https://www.rand.org/pubs/monograph_reports /MR1110.html.

Tey, Tsun Hang. 2008. "Singapore's Electoral System: Government by the People." *Legal Studies* 28, no. 4: 610–28.

Thompson, William R. 1990. "Long Waves, Technological Innovation, and Relative Decline." *International Organization* 44, no. 2: 201–33.

Thurston, Anne F. 1987. *Enemies of the People: The Ordeal of the Intellectuals in China's Great Cultural Revolution.* New York: Knopf.

Times Higher Education. 2022. "Emerging Economies University Rankings." https://www.timeshighereducation.com/world-university-rankings/2022 /emerging-economies-university-rankings#!/page/0/length/25/sort_by /rank/sort_order/asc/cols/stats.

Todaro, Michael P. 1977. *Economic Development in the Third World.* London: Longman.

Tomoshige, Hideki. 2022. "Japan's Semiconductor Industrial Policy from the 1970s to Today." Center for Strategic and International Studies, September 19. https:// www.csis.org/blogs/perspectives-innovation/japans-semiconductor -industrial-policy-1970s-today.

Toner, Helen, Jenny Xiao, and Jeffrey Ding. 2023. "The Illusion of China's AI Progress." *Foreign Affairs*, June 2, 2023. https://www.foreignaffairs.com/china /illusion-chinas-ai-prowess-regulation-helen-toner.

Tong, Yanqi. 1994. "State, Society, and Political Change in China and Hungary." *Comparative Politics* 26, no. 3: 333–53.

Tornell, Aaron. 1995. "Are Economic Crises Necessary for Trade Liberalization and Fiscal Reform? The Mexican Experience." In *Reform, Recovery, and Growth: Latin America and the Middle East*, edited by Rudiger Dornbusch and Sebastian Edwards. Chicago: University of Chicago Press. Accessed April 20, 2018. http:// www.nber.org/chapters/c7650.

Trakman, Leon E., and Nicola W. Ranieri. 2013. "Foreign Direct Investment: A Historical Perspective." In *Regionalism in International Investment Law.* Oxford: Oxford University Press.

Trivedi, Anjani. 2021. "How China's Car Batteries Conquered the World." Bloomberg.com, December 2.

Truex, Rory. 2017. "The Myth of the Democratic Advantage." *Studies in Comparative International Development* 52, no. 3: 261–77.

Truex, Rory. 2019. "Focal Points, Dissident Calendars, and Preemptive Repression." *Journal of Conflict Resolution* 63, no. 4: 1032–52.

Tsai, Kellee S. 1996. "Women and the State in Post-1949 Rural China." *Journal of International Affairs* 49, no. 2: 493–524. http://www.jstor.org/stable/24357569.

Tsai, Kuen-Hung, and Jiann-Chyuan Wang. 2005. "An Examination of Taiwan's Innovation Policy Measures and Their Effects." *International Journal of Technology and Globalization* 1, no. 2: 239–57.

Tsai, Pan-Long. 1999. "Explaining Taiwan's Economic Miracle: Are the Revisionists Right?" *Agenda: A Journal of Policy Analysis and Reform* 6, no. 1: 69–82.

Tsourapas, Gerasimos. 2019. "A Tightening Grip Abroad: Authoritarian Regimes Target Their Emigrant and Diaspora Communities." *Migration Information*

Source, August 22. https://www.migrationpolicy.org/article/authoritarian -regimes-target-their-emigrant-and-diaspora-communities.

Tunsjø, Øystein. 2018. *The Return of Bipolarity in World Politics: China, the United States, and Geostructural Realism*. New York: Columbia University Press.

Uehara, Masashi, and Akira Tanaka. 2020. "China to Overtake US Economy by 2028–29 in COVID's Wake: JCER." *Nikkei Asia*, December 10.

Ugoani, John. 2016. "Education Corruption and Teacher Absenteeism in Nigeria." *Independent Journal of Management & Production* 7, no. 2: 546–66.

Uniacke, Robert. 2022. "Digital Repression for Authoritarian Evolution in Saudi Arabia." In *New Authoritarian Practices in the Middle East and North Africa*, edited by Francesco Cavatorta, Merouan Mekouar, and Ozgun Topak. Edinburgh: Edinburgh University Press.

United Nations. 2022. "Global Flow of Tertiary-Level Students." Accessed September 15, 2022. http://uis.unesco.org/en/uis-student-flow.

"UN Tourism Applauds Saudi Arabia's Historic Milestone of 100 Million Tourists." 2024. *UN Tourism*, February 26. https://www.unwto.org/news/un-tourism -applauds-saudi-arabia-s-historic-milestone-of-100-million-tourist-arrivals.

US Department of Defense. 2021. *China Military Power Report*, XI. https://media .defense.gov/2021/Nov/03/2002885874/-1/-1/0/2021-CMPR-FINAL.PDF.

US Department of State. 2016. *Country Reports on Human Rights Practices for 2016*. https://www.state.gov/reports/2016-country-reports-on-human-rights -practices/.

Vagle, Ben A., and Stephen G. Brooks. 2025. *Command of Commerce: America's Enduring Power Advantage over China*. Oxford: Oxford University Press.

Vallas, S. P., D. Kleinman, and D. Biscotti. 2011. "Political Structures and the Making of US Biotechnology." In *State of Innovation: The US Government's Role in Technology Development*, edited by F. Block and M. Keller. Boulder, CO: Paradigm.

Van de Walle, Nicholas. 2001. *African Economies and the Politics of Permanent Crisis, 1979–1999*. Cambridge: Cambridge University Press.

Vasagar, Jeevan. 2020. "Singapore Should Tear Down its Statue of Raffles." *Nikkei Asia*, June 17.

Vasagar, Jeevan. 2022. *Lion City: Singapore and the Invention of Modern Asia*. New York: Simon and Schuster.

Vermeer, Eduard B. 2006. "Demographic Dimensions of China's Development." *Population and Development Review* 32, no. 1: 115–44.

Vogel, Ezra F. 2011. "Nation Rebuilders: Mustafa Kemal Atatürk, Lee Kuan Yew, Deng Xiaoping, and Park Chung Hee." In *The Park Chung Hee Era: The Transformation of South Korea*, edited by Byung-Kook Kim and Ezra F. Vogel. Cambridge, MA: Harvard University Press.

Vogel, Ezra F. 2013. *Deng Xiaoping and the Transformation of China*. Cambridge, MA: Harvard University Press.

Voo, Julia, Irfan Hemani, Simon Jones, Winnona DeSombre, Daniel Cassidy, and Anina Schwarzenback. 2020. "National Cyber Power Index 2020: Methodology and Analytical Considerations." Belfer Center for Science and International Affairs, Harvard University. https://www.belfercenter.org/sites/default/files /2020–09/NCPI_2020.pdf.

Wade, Robert H. 1990. *Governing the Market: Economic Theory and the Role of Government in East Asian Industrialization*. Ithaca, NY: Cornell University Press.

Wade, Robert H. 2017. "The American Paradox: Ideology of Free Markets and the Hidden Practice of Directional Thrust." *Cambridge Journal of Economics* 41, no. 3: 859–80.

Wakeman, Frederic. 1997. *Strangers at the Gate: Social Disorder in South China, 1839–1861*. Berkeley: University of California Press.

Walder, Andrew G. 2014. "Rebellion and Repression in China, 1966–1971." *Social Science History* 38, no. 3–4: 522–23.

Walder, Andrew G., and Jean C. Oi. 1999. *Property Rights and Economic Reform in China*. Palo Alto, CA: Stanford University Press.

Walder, Andrew G., and Yang Su. 2003. "The Cultural Revolution in the Countryside: Scope, Timing, and Human Impact." *China Quarterly*, no. 173: 82–107.

Wallace, Jeremy L. 2022. *Seeking Truth and Hiding Facts: Information, Ideology, and Authoritarianism in China*. Oxford: Oxford University Press.

Wang, Dan. 2023. "China's Hidden Tech Revolution." *Foreign Affairs* 102, no. 2. Accessed March 22, 2023. https://www.foreignaffairs.com/china/chinas-hidden-tech-revolution-how-beijing-threatens-us-dominance-dan-wang.

Wang, Feng. 2024. *China's Age of Abundance: Origins, Ascendance, and Aftermath*. Cambridge: Cambridge University Press.

Wang, Youqin. 2001. "Student Attacks Against Teachers: The Revolution of 1966." *Issues and Studies* 37, no. 1: 29–79.

Wang, Yuhua. 2015. *Tying the Autocrat's Hands: The Rise of the Rule of Law in China*. Cambridge: Cambridge University Press.

Wang, Dakuo, and Gloria Mark. 2015. "Internet Censorship in China: Examining User Awareness and Attitudes." *ACM Transactions on Computer-Human Interaction (TOCHI)* 22, no. 6: 1–22.

Wang, Di, Tao Zhou, and Mengmeng Wang. 2021. "Information and Communication Technology (ICT), Digital Divide and Urbanization: Evidence from Chinese Cities." *Technology in Society* 64: 101516.

Wang, Feng, and Andrew Mason. 2007. "Demographic Dividend and Prospects for Economic Development in China." United Nations Expert Group Meeting on Social and Economic Implications of Changing Population Structures 141: 141–54.

Wang, Hanchen, Tianfan Fu, Yuanqi Du, Wenhao Gao, Kexin Huang, Ziming Liu, Payal Chandak, et al. 2023. "Scientific Discovery in the Age of Artificial Intelligence." *Nature* 620, no. 7972: 47–60.

Wang, Yuhua, and Carl Minzner. 2015. "The Rise of the Chinese Security State." *China Quarterly*, no. 222: 339–59.

Waterbury, John. 1992. "The Heart of the Matter? Public Enterprise and the Adjustment Process." In *The Politics of Economic Adjustment: International Constraints, Distributive Conflicts and the State*, edited by Stephan Haggard and Robert R. Kaufman. Princeton, NJ: Princeton University Press.

Way, Lucan Ahmad. 2023. "Don't Count the Dictators Out." *Foreign Affairs*, 102, no. 4: 103–115.

Weber, Eugen. 1976. *Peasants into Frenchmen: The Modernization of Rural France, 1870–1914*. Palo Alto, CA: Stanford University Press.

Wedeman, Andrew H. 1997. "Looters, Rent-Scrapers, and Dividend-Collectors: Corruption and Growth in Zaire, South Korea, and the Philippines." *Journal of Developing Areas* 31, no. 4: 457–78.

Wedeman, Andrew H. 2012. *Double Paradox: Rapid Growth and Rising Corruption in China*. Ithaca, NY: Cornell University Press.

Wedeman, Andrew H. 2017. "Xi Jinping's Tiger Hunt: Anti-Corruption Campaign or Factional Purge?" *Modern China Studies* 24, no. 2: 35–94.

Weede, Erich. 1996. "Political Regime Type and Variation in Economic Growth Rates." *Constitutional Political Economy* 7, no. 3: 167–76.

Weeks, Jessica L. P. 2014. *Dictators at War and Peace*. Ithaca, NY: Cornell University Press.

Weinstein, Emily, Channing Lee, Ryan Fedasiuk, and Anna Puglisi. 2022. "China's State Key Laboratory System: A View into China's Innovation System." Center for Security and Emerging Technology, Georgetown University. Accessed August 13, 2024. https://cset.georgetown.edu/publication/chinas-state-key -laboratory-system/.

Weiss, Linda. 2014. *America Inc.?: Innovation and Enterprise in the National Security State*. Ithaca, NY: Cornell University Press.

Wen, Jun, Hadi Hussain, Renai Jiang, and Junaid Waheed. 2023. "Overcoming the Digital Divide with ICT Diffusion: Multivariate and Spatial Analysis at China's Provincial Level." *SAGE Open* 13, no. 1: 21582440231159323.

Wertheim, Stephen. 2020. "The Price of Primacy: Why America Shouldn't Dominate the World." *Foreign Affairs* 99. Accessed August 18, 2021. https:// www.foreignaffairs.com/articles/afghanistan/2020–02–10/price-primacy.

Westad, Odd Arne. 2015. *Restless Empire: China and the World Since 1750*. New York: Basic Books.

White, Gordon. 1994. "Civil Society, Democratization, and Development (I): Clearing the Analytical Ground." *Democratization* 1, no. 2: 375–90.

Whiting, Susan H. 2001. *Power and Wealth in Rural China: The Political Economy of Institutional Change*. Cambridge: Cambridge University Press.

Whyte, Martin King, Wang Feng, and Yong Cai. 2015. "Challenging Myths About China's One-Child Policy." *China Journal* 74: 144–59.

Wiktorowicz, Quintan. 2000. "Civil Society as Social Control: State Power in Jordan." *Comparative Politics* 33, no. 1: 43–61.

Williamson, John, and Stephan Haggard. 1994. "The Political Conditions for Economic Reform." In *The Political Economy of Policy Reform*, edited by John Williamson. Washington DC: Institute for International Economics.

Wilson, Ann-Marie. 2012. "How the Methods Used to Eliminate Foot Binding in China Can Be Employed to Eradicate Female Genital Mutilation." *Journal of Gender Studies* 22, no. 1: 17–37. https://doi.org/10.1080/09589236.2012.681182.

Woertzel, Jonathan, Yougang Chen, James Manyika, Eric Roth, Jeongmin Seong, and Jason Lee. 2015. "The China Effect on Global Innovation." McKinsey Global Institute. https://www.mckinsey.com/~/media/mckinsey/featured%20insights /Innovation/Gauging%20the%20strength%20of%20Chinese%20innovation /MGI%20China%20Effect_Executive%20summary_October_2015.ashx.

Wohlforth, William C. 2020. "Realism and Great Power Subversion." *International Relations* 34, no. 4: 459–81.

Wong, Poh-Kam. 1988. "The Role of State in Singapore's Industrial Development." In *Industrial Policy, Innovation and Economic Growth: The Experience of Japan and the East Asian NIEs*, edited by Poh-Kam Wong and Chee-Yuen Ng. Singapore: Singapore University Press.

Wong, Koi Nyen, and Tuck Cheong Tang. 2010. "Tourism and Openness to Trade in Singapore: Evidence Using Aggregate and Country-Level Data." *Tourism Economics* 16, no. 4: 965–80.

Woo, Wing Thye. 1994. "The Art of Reforming Centrally Planned Economies: Comparing China, Poland, and Russia." *Journal of Comparative Economics* 18, no. 3: 276–308.

Wood, Colleen. 2023. "Between a Rock and a Hard Place: How Kazakhstan's Civil Society Navigates Precarity." *International Labor and Working-Class History* 103: 44–61.

Work, Robert O., and Greg Grant. 2019. "Beating the Americans at Their Own Game." Washington, DC: Center for a New American Security. https://www.cnas.org/publications/reports/beating-the-americans-at-their-own-game.

World Bank Group. 2020. "Promoting Innovation in China: Lessons from International Good Practice." Washington, DC: World Bank. Accessed December 6, 2022. https://openknowledge.worldbank.org/handle/10986/33680.

World Bank Group. 2022. "Government Expenditure on Education, Total (% of GDP)." Accessed March 13, 2025. https://data.worldbank.org/indicator/SE.XPD.TOTL.GD.ZS?locations=CN. Washington DC: World Bank.

World Bank Group. 2023a. "School Enrollment, Tertiary (% Gross)." Accessed March 13, 2025. https://data.worldbank.org/indicator/SE.TER.ENRR?locations=CN&name_desc=false. Washington, DC: World Bank.

World Bank Group. 2023b. "Individuals Using the Internet (% of Population)—Singapore." https://data.worldbank.org/indicator/IT.NET.USER.ZS?locations=SG. Washington DC: World Bank.

World Bank Group. 2024. "Per Capita GDP in Constant 2015 $USD—China." Accessed March 21, 2025. https://data.worldbank.org/indicator/NY.GDP.PCAP.KD?locations=CN. Washington DC: World Bank.

World Bank Group. 2025. "GDP per capita GDP (current $US)." Accessed March 24, 2025. https://data.worldbank.org. Washington DC: World Bank.

Worth, Robert F. 2020. "Mohammed bin Zayed's Dark Vision of the Middle East's Future." *New York Times*, January 9. https://www.nytimes.com/2020/01/09/magazine/united-arab-emirates-mohammed-bin-zayed.html. Worthing, Peter M. 2007. *A Military History of Modern China: From the Manchu Conquest to Tian'anmen Square*. Westport, CT: Praeger Security International.

Wright, Joseph. 2008. "Do Authoritarian Institutions Constrain? How Legislatures Affect Economic Growth and Investment." *American Journal of Political Science* 52, no. 2: 322–43.

Wright, Logan. 2022. "China's Slow-Motion Financial Crisis is Unfolding as Expected." Center for Strategic and International Studies, September 21. https://www.csis.org/analysis/chinas-slow-motion-financial-crisis-unfolding-expected.

Wright, Mary Clabaugh. 1957. *The Last Stand of Chinese Conservatism: The T'ung-Chih Restoration, 1862–1874*. Palo Alto, CA: Stanford University Press.

Wright, Timothy. 2021. "Is China Gliding toward a FOBS Capability?" International Institute of Strategic Studies, October 22. https://www.iiss.org/en/online-analysis/online-analysis/2021/10/is-china-gliding-toward-a-fobs-capability/.

Wu, Friedrich W. Y. 1981. "From Self-Reliance to Interdependence?: Developmental Strategy and Foreign Economic Policy in Post-Mao China." *Modern China* 7, no. 4: 445–82.

Wuthnow, Joel, and Phillip C. Saunders. 2024. *China's Quest for Military Supremacy*. Hoboken NJ: John Wiley, 2024.

Xiao, Gong Qing. 2019. "China's Four Decades of Reforms: A View from Neo-Authoritarianism." *Man and the Economy* 6, no. 1: 1–7.

Xu, Zeyu. 2024. "'Overcapacity' Talks Will Hurt America, Not China." *Sinical China*, April 26. https://www.sinicalchina.com/p/overcapacity-talks-will-hurt-america.

Xu, Beina. 2014. "Media Censorship in China." *CFR Backgrounder*, September 25. Council on Foreign Relations. https://www.files.ethz.ch/isn/177388/media%20censorship%20in%20china.pdf.

Xu, Xu. 2021. "To Repress or to Co-Opt? Authoritarian Control in the Age of Digital Surveillance." *American Journal of Political Science* 65, no. 2: 309–25.

Xu, Chenggang, and Di Guo. 2023. "Is Today's China Yesterday's Soviet Union?" *Japan Times*, January 7. https://www.japantimes.co.jp/opinion/2023/01/07/commentary/world-commentary/china-totalitarianism/.

Yakavets, Natalia, and Makpal Dzhadrina. 2014. "Educational Reform in Kazakhstan: Entering the World Arena." In *Education Reform and Internationalisation: The Case of School Reform in Kazakhstan*, edited by David Bridges. Cambridge: Cambridge University Press.

Yan, Xiaojun, and Jie Huang. 2017. "Navigating Unknown Waters: The Chinese Communist Party's New Presence in the Private Sector." *China Review* 17, no. 2: 37–63.

Yang, Dali L. 1996. *Calamity and Reform in China: State, Rural Society, and Institutional Change Since the Great Leap Famine*. Stanford University Press.

Yang, Su, 2011. *Collective Killings in Rural China during the Cultural Revolution*. Cambridge: Cambridge University Press.

Yang, Tony Zirui. forthcoming. "Normalization of Censorship: Evidence from China." *Journal of Politics*. https://www.journals.uchicago.edu/doi/10.1086/734239.

Yang, R., L. Vidovich, and J. Currie. 2006. "'Dancing in a Cage': Changing Autonomy in Chinese Higher Education." *Higher Education* 54, no. 4: 575–92. https://doi.org/10.1007/s10734–006–9009–5.

Yom, Sean L. 2009. "Jordan: Ten More Years of Autocracy." *Journal of Democracy* 20, no. 4: 151–66. https://doi.org/10.1353/jod.0.0125.

Young, Alwyn. 1995. "The Tyranny of Numbers: Confronting the Statistical Realities of the East Asian Growth Experience." *Quarterly Journal of Economics* 110, no. 3: 641–80.

Yu, Hong. 2014. "The Ascendency of State-Owned Enterprises in China: Development, Controversy and Problems." *Journal of Contemporary China* 23, no. 85: 161–82.

Yu, Hong. 2019. "Reform of State-Owned Enterprises in China: The Chinese Communist Party Strikes Back." *Asian Studies Review* 43, no. 2: 332–51.

Yu, Jie. 2024. "What China Got Right in Latin America." *Foreign Policy*, December 16. https://foreignpolicy.com/2024/12/16/china-latin-america-united-states-development-diplomacy-investment/.

Zee, Josiah. 2011. "Defending Singapore's Internal Security Act: Balancing the Need for National Security with the Rule of Law." *eLaw Journal* 18, no. 1: 28–46.

Zeng, J. 1887. "China. The Sleep and the Awakening," *Asiatic Quarterly Review* 3: 1–10.

Zeng, Jinghan. 2020. "Artificial Intelligence and China's Authoritarian Governance." *International Affairs* 96, no. 6: 1441–59.

Zenglein, Max J., and Anna Holzmann. 2019. "Evolving Made in China 2025." Working Paper No. 8. Mercator Institute for China Studies, Berlin, Germany. https://merics.org/en/report/evolving-made-china-2025.

Zhang, Angela Huyue. 2024. *High Wire: How China Regulates Big Tech and Governs Its Economy*. Oxford: Oxford University Press.

Zhang, Marina Yue. 2023. "Can China Achieve Semiconductor Self-Sufficiency?" *The National Interest*, June 26. Accessed July 18, 2023. https://nationalinterest .org/blog/can-china-achieve-semiconductor-self-sufficiency-206584.

Zhang, Daqing, and Unschuld Paul U. 2008. "China's Barefoot Doctor: Past, Present, and Future." *Lancet* 372, no. 9653: 1865–67. https://doi.org/10.1016 /S0140–6736(08)61355–0.

Zheng, Wang. 2010. "Creating a Socialist Feminist Cultural Front: Women of China (1949–1966)." *China Quarterly*, no. 204: 827–49.

Zheng, Yongnian. 1993. "Nationalism, Neo-Authoritarianism, and Political Liberalism: Are They Shaping Political Agendas in China?" *Asian Affairs* 19, no. 4: 207–27.

Zhou, Kate. 1996. *How The Farmers Changed China: Power of the People*. Boulder, CO: Westview Press.

Zhou, Kate. 2009. *China's Long March to Freedom: Grassroots Modernization*. New Brunswick, NJ: Transaction Publishers.

Zhou, Kai, and Ge Xin. 2020. "Borrowing Wisdom from Abroad: Overseas Training for Political Elite in Reform-Era China." *China Review* 20, no. 4: 95–128.

Zhu, Julie, Fanny Potkin, Eduardo Baptista, and Michael Martina. 2023. "China Quietly Recruits Overseas Chip Talent as US Tightens Curbs." *Reuters*, August 24. Accessed April 22, 2024. https://www.reuters.com/technology/china -quietly-recruits-overseas-chip-talent-us-tightens-curbs-2023–08–24/.

Zhu, Xiaodong. 2012. "Understanding China's Growth: Past, Present, and Future." *Journal of Economic Perspectives* 26, no. 4: 103–24. https://doi.org/10.1257/jep .26.4.103.

Zweig, David, and Stanley Rosen. 2003. "How China Trained a New Generation Abroad." SciDev.Net, August 5. https://www.scidev.net/global/features /how-china-trained-a-new-generation-abroad/.

Zwetsloot, Remco, Jack Corrigan, Emily S. Weinstein, Dahlia Peterson, Diana Gehlhaus, and Ryan Fedasiuk. 2021. "China Is Fast Outpacing U.S. STEM PhD Growth." *CSET Data Brief*. Center for Security and Emerging Technology. https://cset.georgetown.edu/publication/china-is-fast-outpacing-u-s-stem -phd-growth/.

Index